SUN EARTH

How to use solar and climatic energies

by
Richard L. Crowther AIA
Concepts, text

Paul Karius
Graphics, text

Lawrence Atkinson
Text, cover design

Donald J. Frey
Text, editing

of **Crowther** / SolarGroup / Architects

Charles Scribner's Sons
New York

Copyright © 1977 Richard L. Crowther

Library of Congress Cataloging in Publication Data

Crowther, Richard L
    Sun/Earth.

    Bibliography: p.
    Includes index.
    1. Solar energy. 2. Power resources.
3. Buildings—Energy conservation. I. Title.
TJ810.C76 1978   621.47   77-20006
ISBN 0-684-15545-1

This book was first published by Crowther/Solar
Group, Denver, Colorado

Printed in the United States of America

Two drawings in this book have been adapted from Victor Olgyay,
Design With Climate: Bioclimatic Approach to Architectural
Regionalism (copyright 1963 by Princeton University Press):
Figs. 37, p. 16 and 174, p. 89. Reprinted by permission of
Princeton University Press.

foreword

Interesting, imaginative and eminently readable — these are all adjectives which appropriately describe this unique survey of architecture and its relation to the natural environment. Architects are gifted with the ability to illustrate their literary efforts with strikingly effective graphics and handsome lettering. Sun/Earth shows this talent to excellent advantage. Of primary significance, however, is the amount of information which is conveyed, generally in non-technical language, to the reader who is interested in the many ways in which natural forces may be used to produce comfortable living conditions in otherwise uncomfortable situations, thus conserving irreplaceable stores of the fossil fuels which the sun produced so many million years ago, and which are not, to the best of our knowledge, continuing to be produced now. We must learn to live with our environments and Sun/Earth will help in this endeavor.

John I. Yellott

John I. Yellott, Professional Engineer
Visiting Professor in Architecture
Arizona State University
Tempe, Arizona

Acknowledgements

The author wishes to express special recognition and appreciation to Donald Frey P.E. for his outstanding ability in providing organization, clarity, and technical substance to the text, to Paul Karius for organization, research, and graphics, and to Lawrence Atkinson for the cover design and general assistance.

The author also gratefully acknowledges the personal time, suggestions, and critical comments kindly given by the following persons during the preparation of Sun/Earth.

John I. Yellott, Professional Engineer
Visiting Professor in Architecture
Arizona State University
Tempe, Arizona

Benjamin T. (Buck) Rogers
Los Alamos Scientific Laboratory
University of California
Los Alamos, New Mexico

DeVon Carlson, FAIA
Professor of Architecture
University of Colorado
Boulder, Colorado

Jerry D. Plunkett, Ph.D.
The Montana Energy and MHD Research and
    Development Institute, Inc.
Butte, Montana

Frank R. Eldridge
The Mitre Corporation
McLean, Virginia

Elizabeth Kingman
Solar Energy Exhibition Program
University of Colorado
Denver, Colorado

Crowther/Solar Group takes full responsibility for the inclusion of various concepts, descriptions, content, and accuracy of the Sun/Earth material. No endorsements by the above mentioned people are implied or were given.

Appreciation for special assistance is extended to Diana Gilmore, Cheryl Amidon, and others who gave their time and talents to the preparation of this book.

table of contents

biographical comment

To find Richard Crowther as the author of a publication having this scope and content is no surprise to those of us who are familiar with his multi-faceted career. He has concern for effective innovation, for assisting us to an improved state of living, and for stimulating us to thought about the future — all of this is characteristic of him and his work.

Architecture and planning, to him, is an inventive science, the product of which is a composition of building and site which year—round responds effectively to seasonal climatic conditions. Guided by these philosophic concerns, he early called for energy conservation and, in architecture and planning, a response to solar and other direct natural energies. He has prophetically led us through social and technological change to, finally, holistic creative design.

In the early thirties and forties, he was a pioneer in the developing contemporary architecture. This pioneering quality still guides his works and service as they encompass not only architecture but also interior design, research, and land and topographic planning for commercial, institutional, and residential purposes.

The effectiveness of this designer has been extended through consulting appointments by federal, state, and city agencies, as well as private organizations. His professional colleagues have sought his advice and assistance in programs promulgated by the American Institute of Architects (AIA) at both local and national levels. He serves as a member of the Task Force for the "Energy Notebook" of the AIA Research Corporation.

Through invited lectures to environmental, political, business, and social organizations he has stimulated others to awareness and action for a better environment and for more efficient use of resources. He has served as lecturer at the Universities of Colorado and Wisconsin and the Midwest Research Institute. His contributions to various publications have reached those outside his professional circles; among these have been articles in the National Geographic, the Journal of the American Society of Heating and Ventilating Engineers, House Beautiful Building Manual, and various environmental publications.

photo by Malcolm Wells

Theories espoused herein have been converted, over many years, into real aids to improved living through his service to clients and through projects which he himself has inaugurated. Currently over twenty residential or commercial projects applying these principles are at the planning, construction, or recently completed stages. The latest example is the new office for Crowther/Solar Group which employs optimal solar energy conservation for their activities and for related educational facilities. All solar commercial and residential projects use optimized energy conservation, "passive" energy systems, and "active" solar energy collection systems.

This volume is his latest contribution to increased understanding and more effective use of natural energies.

DeVon M. Carlson FAIA
Professor of Architecture
University of Colorado
Boulder, Colorado

introduction

The forces of nature are regenerative, deriving their impetus from the sun and universe. To ensure man's survival, the fragmented, isolated systems and constructions of man need to acquire the elegant holistic interdependence of nature's resources and systems. Nature remains as the most sophisticated designer of the cosmic and world environment. Man's systems need to be made fully regenerative in concert with natural forces and nature's bounteous provisions.

This book graphically presents many concepts which are cost effective today for the utilization of free natural energy sources in homes and other buildings. All of the natural energy concepts presented are in a process of continuing development. Many of them are immediately economic and practical, while some are not.

It takes the application of money to construct devices to harness natural energy or to construct energy efficient forms of architecture. In numerous cases operational energy is not required to employ the sun, wind, water, and earth as free anti-inflationary energy sources. In other cases a very small input of operational energy in comparison to the total energy output is required.

All land and buildings are solar collectors. The problem is how to cost effectively make them efficient collectors of solar radiation in winter and how to use natural forms of energy to cool and ventilate them during summer and other seasons of the year.

Regional and microclimatic conditions vary throughout the world. Topography and landscaping can play an important role in climatic control and climatic effect upon architecture. The examples presented for optimized energy conservation and solar active and passive systems are generic to most northern latitudes, but need modification or adaption to specific locations and climates.

An annotated bibliography, containing additional references, is included for those individuals who wish to go beyond the presentations of this book. Also included is an appendix of energy related data.

Richard L. Crowther AIA

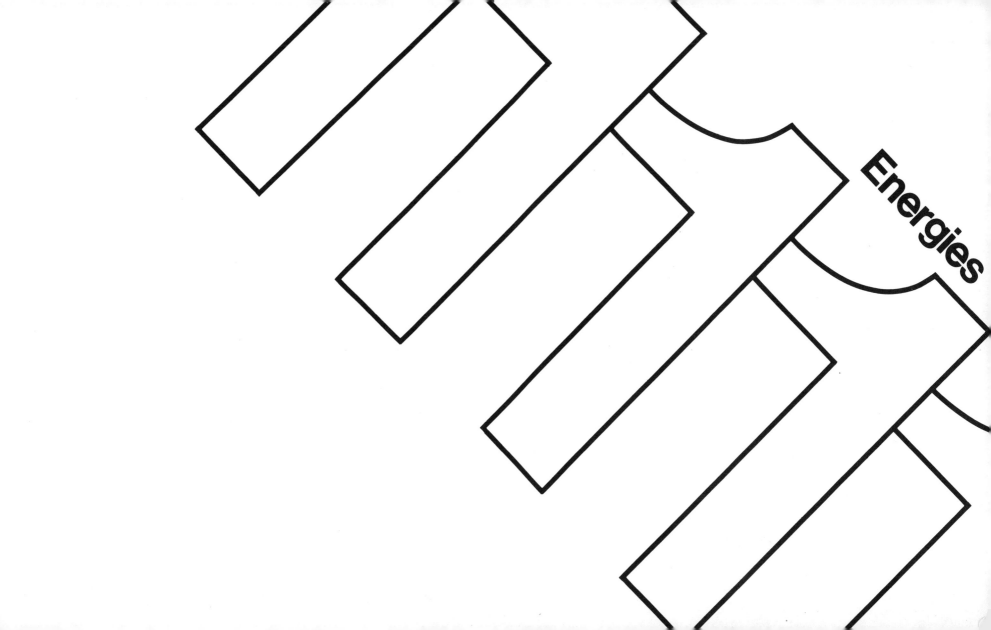

Energies

Technological man has enjoyed the luxury of abundant energy and fuel supplies for many years. Now these reserves, which are finite or limited in quantity, have diminished to a point where primary reliance upon them is no longer justified. Millions of years were necessary for their formation, and because they are being used at increasing rates, their quantity is rapidly decreasing. The majority of this consumption has occurred during the two centuries following the industrial revolution. The energy which is now consumed in one day, whether oil, natural gas, or coal, originally took thousands of years to form. The technology does not exist to extract completely the energy potential from the fossil fuels used; their incomplete combustion is, in fact, a major contributor to air pollution. Developing countries are steadily increasing their demand for fossil fuels. These demands, added to present rates of consumption, will contribute to severe energy shortages in the future.

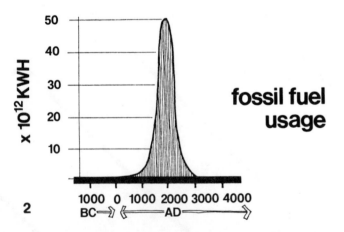

fossil fuel usage

Man has traditionally attempted to dominate the environment instead of harmonizing with it. He has used his intelligence to try to subdue nature and resist the natural forces of the universe. Playing the role of a superior being, he is destroying the earth as he extends his influence and accelerates his technology. Since the Stone Age, man has modified his physical surroundings for protection and comfort, and sometimes merely as a demonstration of his control over nature. As he developed the use of tools, he decreased the proportion of direct physical energy needed to complete a task. Early man sought to amplify his own energy by means of simple machines such as levers, pulleys, and various forms of inclined planes. Fuel (wood) combustion, meanwhile, was used only in producing heat. As his development continued, man harnessed the power of animals, slaves, and the more complex machines they drove, minimizing his own energy input. This "biological" power source was ultimately superseded by the combustion of fuels to perform mechanical work. Our contemporary use of fossil fuels now includes a third service, the creation of synthetic materials. Among competing applications, these fuels should be allocated to products of long term durability, to the production of medicines, and to other uses of immediate importance which cannot be met in other ways.

Contributing to the present "energy crisis" is society's dependence on combustion heat derived from fossil fuels instead of the almost limitless energy available from the sun. Man must realize that he can benefit directly and indirectly from the natural energy which is available to him.

2

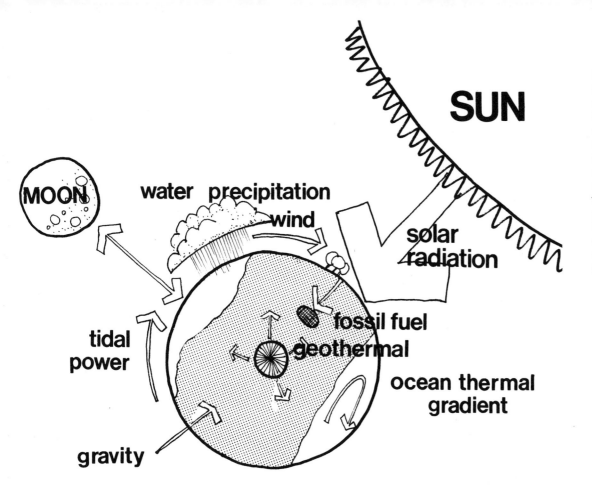

SUN

MOON

water precipitation

wind

solar radiation

tidal power

fossil fuel

geothermal

ocean thermal gradient

gravity

Most of the energy on the earth comes or has come indirectly from the sun. The direct solar energy which strikes this planet (incident solar radiation has evolved the term "insolation") provides and sustains the biosphere in which we live. The radiation from the sun heats our atmosphere and causes air movement or wind. It also heats the oceans and water reservoirs of the earth, and in combination with wind creates weather and sustains the continued water cycle of the planet. The sun provides the energy for plant metabolic activities. Energy is stored by them in the form of carbohydrates, wood, or as fossil fuel. Fossil fuels are created with energy stored by the forces of heat and pressure acting upon decaying plant material.

3

The amount of energy consumed by technological man
is phenomenally large in comparison to that used by
primitive man, who consumed only the energy he
received from his food. His heat was a product of
his own metabolic activity. As man advanced he
began to cook food and warm his dwelling with the
heat from a fire. Agricultural man initiated the use
of tools for farming and the use of animals for labor.
These were the first industrial uses of energy.
Industrial and technological man increased the use of
energy in all categories. This is especially true in
the area of transportation, as a result of man's
dependence on the automobile.

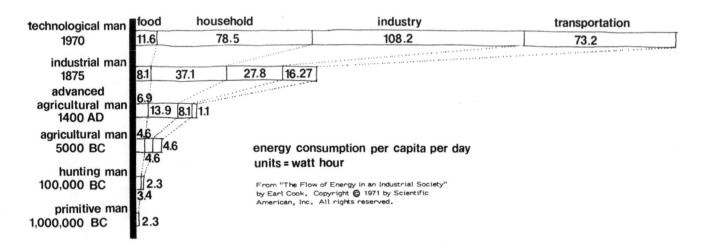

energy consumption per capita per day
units = watt hour

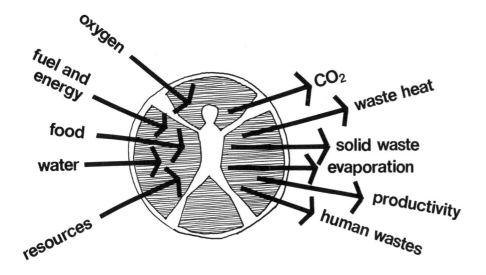

oxygen

fuel and energy

food

water

resources

CO₂

waste heat

solid waste

evaporation

productivity

human wastes

man's inputs and outputs

Numerous inputs and outputs from man's biosphere affect the amount of energy he uses. He has direct uses for such things as heating, cooking, lighting, and transportation, and indirect uses such as the manufacture of goods, use of resources, production of food, and handling of wastes.

Inputs to the biosphere are converted by many processes to outputs. The outputs from the consumption of fuel and energy are solid wastes (particulates and pollution in the air), carbon dioxide (a by-product of combustion), and waste heat (heat produced by combustion exhausted to the environment). Oxygen which is used for respiration is converted to carbon dioxide. Plants take in carbon dioxide and water and produce oxygen during photosynthesis, thus maintaining the oxygen content in the atmosphere. The input of food and water results in an output of solid wastes, human wastes, and heat wastes. The body of an average adult human male gives off a minimum of 250 Btu per hour (Btuh) by convection, radiation, and evaporation while at rest.

The processes involved in the production and distribution of consumer goods require large amounts of energy. This energy is consumed during the extraction and transportation of resources, and during manufacturing processes. It is needed for packaging final products and transporting them to their destinations. Additional energy is needed to purchase any product and transport it to its place of use. The numerous and complicated processes along the way produce waste heat, solid wastes (disposing of these is a major problem in many large cities), air pollution, and carbon dioxide.

The USA, with only six percent of the world's population, uses thirty-three percent of the world's energy and produces over fifty percent of its solid wastes.

As leaders in world technology, and in per capita general resource consumption, we in the United States should be developing systems which use the sun and other natural energies to a much greater extent. We should also examine our individual energy use patterns in attempting an immediate reduction in the massive energy waste of our society.

When we consume fossil fuels to produce energy we must recognize the source of the energy and the by-products of its use.

Plants store the sun's energy as they grow. Certain plants are used as a source of energy by releasing sun-stored energy in a useful form. Burning is a means of releasing the energy of wood, dried grass, or any other combustible material. As a log is burned, the energy which a tree received from the sun and stored while growing is released. The burning process does not convert all the stored energy to usable heat. Some of the wood is not completely burned and escapes as soot and smoke. When wood is completely burned, it releases only carbon dioxide and water vapor (wood is a carbon compound).

Some plant materials and other organic matter are converted to fossil fuels. This process takes many thousands of years and requires proper conditions of heat and pressure.

Deposits of the fossil fuels (oil, natural gas, and coal) are formed only by unusual geological circumstances. In general, the fossil fuels are derived from the remains of land-dwelling plants and other organic matter. Younger sedimentary strata cover the plant remains, and the extreme pressure and temperature produced by succeeding layers are responsible for the transition from plant compounds to the simpler higher energy level hydrocarbons.

Fossil fuels are rarely found in their purest form. Minerals such as sulfur are considered impurities in fossil fuels, and are released as oxides during combustion. The introduction of clay and sand which can occur during coal formation will remain as ash after the coal is burned, incurring a substantial cost for its removal.

The combustion of fossil fuels adds the following to the atmosphere:

a. carbon dioxide (natural product of combustion)
b. carbon monoxide (product of incomplete combustion)
c. water vapor (natural product of combustion)
d. solid particles (non-combusted matter carried as smoke and soot)
e. hydrocarbons (product of incomplete combustion)
f. sulfur, nitrogen, and mineral oxides (resulting from the combustion process)
g. waste heat (from burning and mechanical sources)

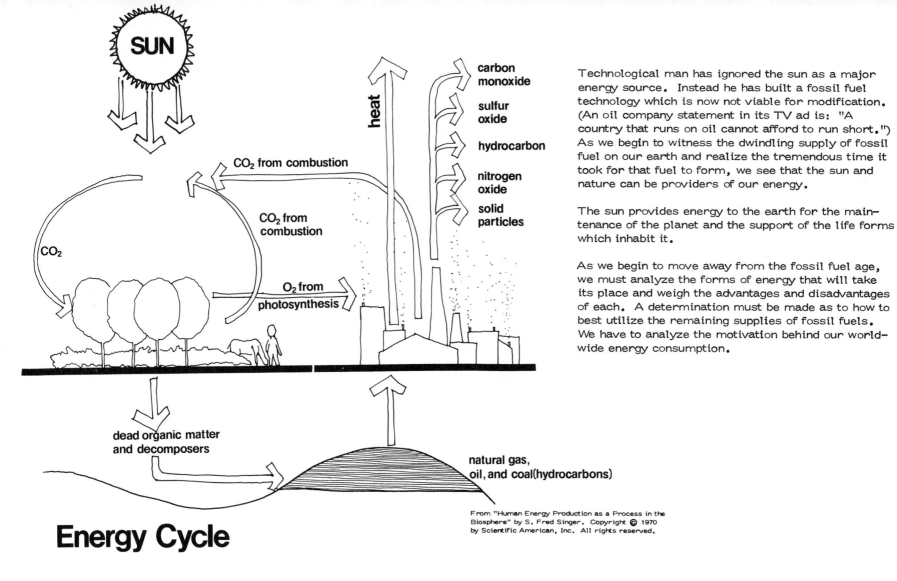

SUN

heat

carbon
monoxide

sulfur
oxide

hydrocarbon

nitrogen
oxide

solid
particles

CO_2 from combustion

CO_2 from
combustion

O_2 from
photosynthesis

CO_2

dead organic matter
and decomposers

natural gas,
oil, and coal(hydrocarbons)

Energy Cycle

Technological man has ignored the sun as a major
energy source. Instead he has built a fossil fuel
technology which is now not viable for modification.
(An oil company statement in its TV ad is: "A
country that runs on oil cannot afford to run short.")
As we begin to witness the dwindling supply of fossil
fuel on our earth and realize the tremendous time it
took for that fuel to form, we see that the sun and
nature can be providers of our energy.

The sun provides energy to the earth for the maintenance of the planet and the support of the life forms
which inhabit it.

As we begin to move away from the fossil fuel age,
we must analyze the forms of energy that will take
its place and weigh the advantages and disadvantages
of each. A determination must be made as to how to
best utilize the remaining supplies of fossil fuels.
We have to analyze the motivation behind our worldwide energy consumption.

All life processes depend on the ability of plants to capture the sun's energy. Green plants use sunlight to build sugar and other compounds from carbon dioxide and water. All other forms of life use this primary energy source in some form.

Animals and man derive all or part of their energy from plants, breaking down and restructuring plant compounds into forms useful to their own metabolism for the extraction of energy. This energy is then converted into bio-mechanical energy and heat as man and animals go about their daily activities.

Some animals, including man, also utilize the energy locked into animal compounds. Again the process of restructuring takes place with reduced efficiency, such as when vegetable proteins are subsumed into meat.

It becomes apparent when viewed in this cyclic pattern, that the most efficient energy use is the one most directly linked to the source of energy: the sun; and that as we proceed through the secondary and tertiary recycling of this energy base, the efficiency is accordingly diminished.

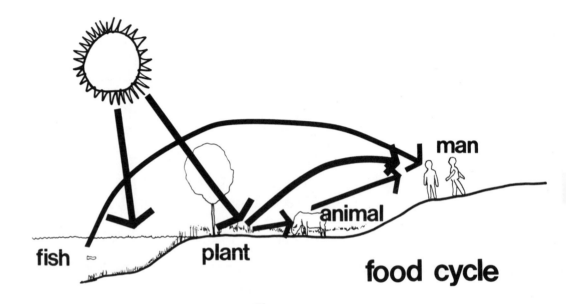

man

fish

plant

animal

food cycle

fossil fuels

The three most universally used fossil fuels are hydrocarbons: coal, oil, and natural gas. Each contains mainly the two elements hydrogen and carbon. Statistics about world reserves and consumption of these fuels must be in terms of their conversion to other forms such as mechanical power, electricity, and synthetic materials, in addition to their familiar use in direct combustion for heat.

Because public utility companies are appropriating these three fossil fuels toward generating electricity, they are becoming less available for direct use in buildings. Unfortunately, thermal energy is "lost" in the form of waste heat at each stage of the conversion from one form of energy to another. Natural gas burned in a furnace delivers less than 70% of its stored energy as heat to a home. If the same natural gas is first converted to electricity at a power generation station, and then the electricity is converted to heat in the home, only 30% of the energy stored in the gas is delivered as heat. During this process, up to 70% of the stored energy is lost as waste heat. Other fossil fuels have even lower conversion efficiency.

The above facts indicate why solar energy, wind energy, and other natural energy forms are being more seriously considered as power sources. Most of these direct sources involve very short, efficient chains of conversion from initial form to final use.

More important, the fossil fuels extracted from the earth are not renewable and are limited in supply, while the supply of energy from the sun is practically unlimited.

coal

Coal is used to supply the United States with approximately one-fifth of its energy needs. However, the United States' reserves of coal in terms of energy potential are actually far more abundant than the reserves of oil or natural gas. Coal is an advanced-stage hydrocarbon, and is the highest density naturally occurring solid fuel. This indicates a lengthy formation period, probably hundreds of millions of years. The duration of this process clearly shows why "mines bear no second crop". Although coal is not renewable, it has become prominent as a fuel because:

a. It burns at a higher temperature and is more compact than wood.

b. Its initial abundance has, so far, outweighed its non-renewability. Eighty-eight percent of all proven U.S. fuel reserves are coal.

c. It can be recovered using several different methods and can be either burned directly or converted to other forms prior to combustion.

The disadvantages inherent in coal have recently become more evident, partly due to increased concern for coal miners' health and safety, and concern due air pollution. Problems facing continued use of coal include:

a. The safety of mining personnel
Technical difficulties in deep-mining coal and accumulations of explosive gases ensure that mine cave-ins will always carry a certain risk. Mining of coal can cause several serious respiratory ailments for miners exposed to coal dust.

b. Impact upon the environment
The alternative method of extraction, strip mining, is beset with public disapproval for environmental reasons.

c. Alternate uses of coal
Coal is being used more frequently for the production of synthetic materials as well as for combustible fuel. This is partly to compensate for oil shortages in the same areas of use. Such trends are expected to accelerate.

d. The involvement of other resources
Elaborate processing of coal involves vast quantities of water which may be needed elsewhere.

e. Alternative uses of the land
Much of the coal reserves are located under arable land which must either be sacrificed or restored at great cost after coal extraction.

f. Reserves that can't be recovered
The above factors may render large percentages of U.S. coal reserves unrecoverable.

g. Potential air pollution
Coal varies widely in heating value per unit weight and in ash and sulfur content. The products of coal combustion require intensive capital investment to reduce atmospheric pollution to acceptable levels.

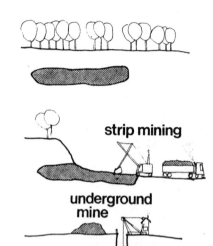

strip mining

underground mine

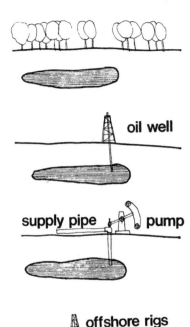

oil well

supply pipe   pump

offshore rigs

h.   Low grade coal
Coal which has a high level of ash and moisture content offers very low heating value per unit weight.

i.   High transportation costs
Coal of low heating value cannot be shipped over a wide radius without transportation cost exceeding the worth of the fuel itself. Certain public utility companies now pay shipping costs of five times their coal material costs.

Despite the negative and unhealthful aspects of burning coal to produce power, it remains as the most immediate, logical, abundant source for electrical generation. Every effort is needed to decrease our escalation of national energy demand, and to supplant coal and other environmentally degrading fuels with clean, natural energy alternatives.

oil

Oil is a highly versatile fuel, supplying over 36% of the total energy requirements of the United States. Because of the extremely slow creation cycle for this hydrocarbon, each year's worldwide oil use represents millions of years original "production" time. This indicates a very rapid depletion of reserves. Increasing problems in its recovery, refining, political influence, and distribution include:

a.   Sudden oil "blowouts", originally termed "gushers" from wells, are now controllable during land-based drilling, but are more dangerous in offshore operations.

b.   Routine refinery operations have produced a greater number of river and ocean "spills" each year, as a natural consequence of increased volume and cumbersome methods of transport. In 1969, 2.2 million tons were spilled into the world's oceans.

c.   Although much of the United States' land area is leased for oil and gas exploration, yields from new discoveries began decreasing in 1970, and the United States' share of known world petroleum reserves had already dropped from 40% in 1937 to 6.7% in 1968. As little as twenty years of recoverable domestic reserves now remain.

d.   As western world dependence on offshore oil begins to overtake land-drilled oil use, the net energy value obtained will decrease drastically. The amount of energy required to extract oil may approach the level of the energy extracted.

The use of oil obtained from unconventional sources such as tar sands, oil shale, and coal liquefaction appears to involve net energy disadvantages similar to those noted above.

Despite the negative and polluting effects of oil as a fuel, it is the only present abundant energy source that can adequately serve transportation needs. Its use for heating should be curtailed, and its reserves extended, possibly by adding ethyl alchohol to gasoline or by powering automobiles with methanol, a derivative of coal.

11

natural gas

Natural gas now fulfills approximately one-third of U.S. energy demand. Considering the economics of pipeline distribution, gas is the most efficient direct-combustion hydrocarbon fuel now in general use. It is also the most cleanly burning, and is the least polluting of the fossil fuels.

The fundamental drawback to the use of this excellent fuel is its limited quantity. This situation and the artificial price controls have led to various problems which are becoming more critical:

a. Natural gas prices in interstate commerce have been held low by federal controls. This has discouraged companies from exploring for it.

b. After intensive exploration and drilling are resumed, the United States will be facing total depletion of natural gas in twenty to fifty years. Imported and synthetic gas will then be the only fuel available in this clean form. In the meantime, prices will rise rapidly to levels approaching those of electricity.

c. Recent land-based natural gas finds have exceeded 20,000 feet in depth, as compared with 10,000 feet prior to 1970. Most authorities agree that the cost of drilling doubles for each 3600 feet of depth, signaling an eightfold drilling cost increase in the past five years alone.

d. As a consequence of the above trends, some gas is now being transported from other continents, primarily Africa. The gas must be in a compact, liquid form during transportation, which is accomplished by refrigeration. Liquid Natural Gas (LNG) ocean tankers carry more potentially destructive power aboard than any other energy "container" except nuclear reactors. In 1971 alone, more than 100 tanker collisions occurred. None of these happened to involve LNG cargo, but the probability will increase as more tankers are put into service.

The use of natural gas as a combustible fuel, or in any other form, faces imminent decline. In many locations no new gas connections are available and industrial and institutional users face the probability of reductions or termination of supplies. Natural gas is an exceptionally valuable raw material from which a host of products are made, ranging from refrigerants to fabrics. This will accelerate a shift in emphasis by governmental agencies toward the various forms of available solar and natural energy.

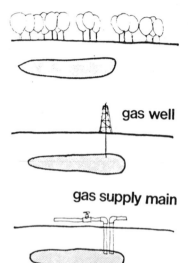

gas well

gas supply main

nuclear fission

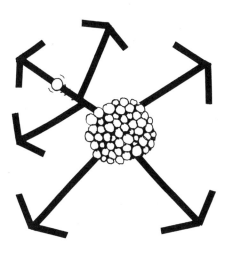

All nuclear reactor power stations presently in operation use fission processes in which heavy uranium (235) atoms are split to produce heat for steam-electricity generation. This is the fastest growing method of power generation in the United States. In 1974 it represented less than 4% of the nation's electrical generating capacity, and 1% of its overall energy supply; this was barely equivalent to the heat energy obtained by burning wood in home fireplaces that same year.

Despite heavy government funding of research and the fact that one "fission event" (with a uranium 235 atom) releases 50 million times as much energy as the burning of a hydrocarbon (such as coal) atom, the development of nuclear energy has not yet approached its projected potential.

Drawbacks to the expansion of the nuclear industry are numerous, major, and often cited but little understood. The most publicly debated aspect of such energy is its safety. Indeed, most of the reactor stations operative in the United States have at one time or another been shut down for a range of reasons from physical failures to design flaws. The following are factors contributing to the drawbacks of nuclear energy.

a. Reactors must be kept cooler than fossil fuel powered plants in order to protect uranium fuel rods. Consequently, their waste heat totals 70% of all the energy released, and thermal pollution is far greater than for any fossil fuel plant.

b. Radioactive waste products, especially the plutonium from fast-breeder fission reactors, are potentially lethal for thousands of years. The complexity and cost of a waste disposal system which insures the safety of the public and the environment are major factors which the industry may not be able to overcome.

c. Uranium fuel comprises less than 2 parts per million of the earth's crust. Unless fission technology shifts to fast-breeder reactors from conventional uranium reactors, uranium supplies may be exhausted within the century. Liquid-metal fast breeder reactor technology is still more complex and creates still more dangerous waste products than do the more conventional fission processes.

d. Conventional industrial "economies of scale" have so far worked in reverse with nuclear energy. Maintenance problems are compounded by safety precautions with every increase in size. Thus, while the fission reactors powering submarines and aircraft carriers operate faithfully and with impressive efficiency, central utility plants for urban electricity are plagued by repairs taking up to one year apiece. These plants must be completely inoperative for any major maintenance or repair; most operate on line only 60% of the time and then function at only 50% to 60% of their rated capacity.

e. Initial capital cost for a nuclear/electric power station is considerably higher than for any other equivalent-scale fossil fuel facility.

f. Licensing, planning, and construction of nuclear plants now in service have ranged from seven to seventeen years, averaging ten years. This greatly exceeds construction time for all other existing energy production plants, except perhaps hydro-electric dams.

g. The fossil fuel quantities expended in the mining of uranium, its processing, and its ultimate disposal in other forms of by-products even more lethal, are impressive. It could lead to the irony that the amount of hydrocarbon fuels consumed to support all nuclear activities would be greater than the electrical power produced by nuclear facilities.

h. There is more energy from the sun striking the land area required for a nuclear power plant and its surrounding area than is produced by the plant itself.

i. The cost of uranium, extracted from minerals, has gone up faster than the cost of Arabian oil.

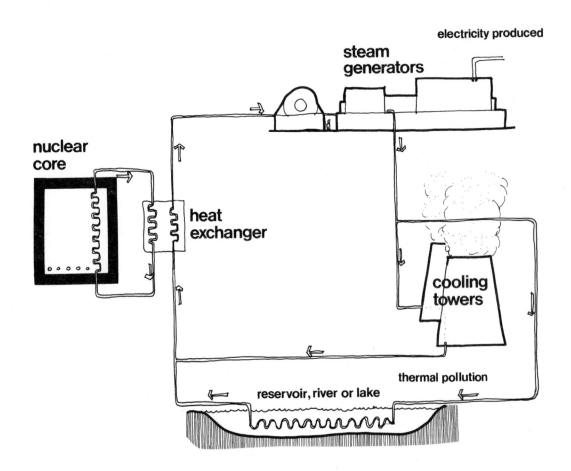

nuclear core

heat exchanger

steam generators

electricity produced

cooling towers

thermal pollution

reservoir, river or lake

nuclear fusion

Thermonuclear fusion as energy potential is theoretically far more abundant than any form of fission, and possibly less dangerous. It has, however, shown itself even more difficult to master than fission and may induce radioactivity in the materials of construction of the generating plant.

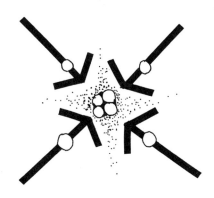

In principle, fusion proceeds by the forceful combining or fusing of very light (hydrogen) atoms. Since hydrogen is the most abundant element in the universe, and the sun continues successful operation as a fusion reactor, such application on earth has become most attractive to scientists. The sun, however, relies on intense gravitational forces as well as high temperatures for hydrogen fusion to take place deep in its core. With such gravity or its simulation unattainable on earth, temperature for fusion must compensate by being still higher than at the sun's core. Magnetic confinement fusion reactors and, more recently, laser fusion reactors have each approached for an instant the required temperature range between 60 and 100 million degrees Celsius; but after twenty years' research in several countries (at least on magnetic confinement fusion) such reactions remain highly undependable. Potential radiation hazards as a result of using heavy hydrogen (deuterium and tritium) are much smaller than those which accompany the use of nuclear fuels; nevertheless, they must be dealt with.

One little-known problem presented by fusion technology is its present requirement for exotic metals, such as vanadium, niobium, and lithium, in processing and heat resistance.

Generally, both fission and fusion nuclear energy have met deepening doubt in the public mind, if not in the minds of the scientific and political communities. Fission is looked upon with due apprehension for its dangers, which have overshadowed an equally severe cost efficiency disadvantage and a limited reserve of fissionable materials.

Breeder reactors raise the problem of the dangers inherent in their principal product, plutonium. Fusion has yet to be accomplished and there is no certainty that it will ever be practically achieved on earth. There is a certain futility in attempts to duplicate the fusion process of the sun, which already delivers its products to earth in a worthwhile form.

Energy comes in many forms. There are also two types of energy – potential and kinetic. Potential energy is stored energy which is available for release. A rock at an elevated point has stored energy. As it is released from that point, its potential energy is liberated. Kinetic energy, the second type, is moving or dynamic energy. As the rock falls, it converts its potential energy into kinetic energy.

The forms energy takes can be divided into five categories: radiant, electrical, chemical, mechanical, and heat. Energy cannot be created or destroyed, but one form of energy can be transformed or converted into another energy form. Heat is the final form of the energy used in many processes. The difficulty or ease of the conversion from one form to another influences the cost of energy.

The more difficult it is to convert energy to useful work or heat, the more expensive and less practical it becomes. An example of this is the automobile. In its internal combustion engine, the potential chemical energy stored in gasoline is converted to mechanical energy (kinetic energy) and heat. The mechanical energy is also converted into heat by the friction of air, tires, and bearings. There are many energy losses before the potential energy stored in the gasoline is converted to kinetic energy (velocity) of the automobile.

Below are some examples of each energy form:

Radiant energy
    sunlight
    fire (conversion from heat energy)
    radiant heaters
    light bulbs (conversion from electrical energy)
    fireflies (conversion from chemical energy)

potential energy

kinetic energy

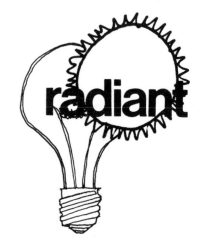

radiant

 latent heat

 sensible heat

ible heat or
t a body gives
se, and is
e. Latent heat
nge its state or

ance is a change
ce may exist in

r air)
y)

only one state
s and pressures
ple of this is
he gaseous state.
F below zero,
quid. (Throughout
s not used. This is
practice.)

substance from a
gas. It is given
. Under stan-
ill increase in
Once a tempera-
heat does not
the impetus for
at the same
Per unit weight
than water at

Water gives off only sensible heat when the temperature range is between boiling and freezing. So if
water at one atmosphere of pressure is cooled from
190 F to 50 F, the heat removed or given off is
sensible heat. If water is cooled from 35 F to 31 F,
the heat removed is a combination of sensible and
latent heat, since water freezes at 32 F.

All substances have the ability to store heat. Some
can store more heat per unit weight than others; the
term for this capacity is specific heat. Temperature
is a measure of the kinetic energy of the molecules of
a substance and is an indicator of the amount of heat
stored in it.

The land, sea, and air share in thermal interaction
and each influences the temperature of the others.
The temperature of the earth's surface determines
the temperature of the atmosphere above it. The
atmosphere in turn helps to maintain and equalize
the temperature of the earth. A coastal city experiences less extreme seasonal and day/night
temperature differences than its inland counterpart. This is due to the relative thermal stability
of water compared to air and land. Such stability
is a result of the high specific heat of water.

Chemical energy
　　plants, food, wood (conversion from solar energy
　　　　by photosynthesis)
　　man's energy (conversion from other chemical
　　　　energy)
　　fossil fuels (long-term conversion and storage
　　　　of radiant energy)
　　battery (conversion to and from electrical energy)

 chemical

Electrical energy
　　electricity (can be readily converted to any of
　　　　the others)
　　static electricity (conversion from mechanical
　　　　energy)

 electrical

Mechanical energy
　　internal and external combustion engine (conver-
　　　　sion from chemical energy)

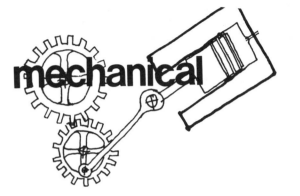 mechanical

Heat energy
　　fire (conversion from chemical energy)
　　heating coils (conversion from electricity or
　　　　geothermal energy)
　　waste heat (by-product of conversion of all forms
　　　　of energy)

heat

There are three ways of transmitting heat energy: conduction, radiation, and convection.

Conduction is the transmission of energy between two bodies which are in direct contact. As a common example, a teapot when placed on an electric stove receives energy by conduction (or direct contact).

Radiation is the transmission of heat by electromagnetic rays. Only these rays can travel through a vacuum such as outer space and heat the object which intercepts them. The sun warms the earth by radiation which travels in this manner.

Convection is the transmission of heat through a fluid. An object heats the air (or liquid) in contact with it, and the warmed fluid travels to nearby objects and warms them. The use of convection for heating is exemplified by a home forced air furnace. The furnace warms the air, then the hot air moves to heat occupants and the house.

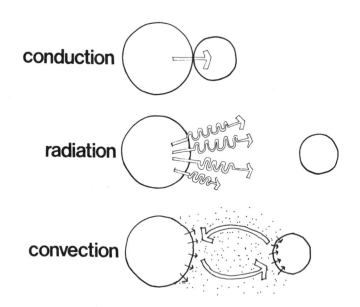

conduction

radiation

convection

energy content

hot

cold

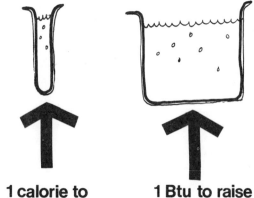

1 calorie to raise 1 gram water 1° C.

1 Btu to raise 1 pound of water 1 degree F.

1 kilowatt-hr. = 3414 Btu

1 calorie = .003968 Btu

1 Btu = 1055.87 joules

A change in tempe
in the energy cont
usually in the form
in a chemical form
signals an increas
in temperature in
tent. Some of the
ing energy are pr

Three units which
Calorie, the Btu
A Calorie is the a
gram of water 1°
required to raise
is the unit of ener
Systeme Internati
the S.I.

* The British no
used in the USA.

The moon has vast temperature differentials and is an example of a planet without an atmosphere. The bright side of the moon (the side which is in direct solar radiation) heats up to several hundred degrees C; the dark side is hundreds of degrees below zero C.

The atmospheric temperature on the earth varies seasonally and with latitude, but doesn't reach the extremes experienced by the moon. The warmth of the atmosphere varies in proportion to the amount of direct solar radiation (insolation) reaching the earth's surface. The water content or humidity greatly affects the air temperature that is perceived. Latent heat is required to vaporize water and add humidity to the air.

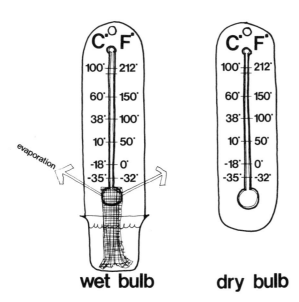

wet bulb   **dry bulb**

The influence of humidity on temperature can be quantified by comparing temperatures measured in two different ways. These are called dry bulb and wet bulb temperature. Dry bulb temperature is a measure which doesn't take into consideration the humidity. This is the method of measuring temperature with which the majority of people are most familiar. The units of temperature are either degrees Celsius, C (often called centigrade), or degrees Fahrenheit, F. The Celsius scale will soon be implemented in the United States as part of a phased program to adopt the S.I. (metric) system of measurements. (The S.I. units will be used in this book, with the more familiar units in parentheses.)

The temperature measured by a wet bulb takes into account the humidity in the air, by measuring the rate of cooling through evaporation at the bulb of the thermometer. When the humidity is low, the wet bulb temperature will differ by many degrees from the dry bulb temperature. As an example, in Arizona during the summer, the dry bulb temperature (the temperature which is announced by a weatherman) may be 42 C (108 F) while the wet bulb temperature may be 21 C (70 F). When the humidity is high the wet bulb temperature will tend to differ only slightly from the dry bulb temperature. St. Louis on a hot muggy day may have a 21 C (70 F) dry bulb temperature and a 20 C (68 F) wet bulb temperature, indicating that the humidity is very high.

The difference between wet bulb and dry bulb temperatures will help explain the use of natural cooling and heating systems in different climates.

The energy relationship between the sun and man's biosphere, and other heat sources outside the biosphere, includes many factors which help to maintain the heat balance of the human organism.

Man can be a receptor of direct or reflected radiation from the sun. He also generates heat from his body processes. The amount of heat he generates depends on his activity level (whether he is sleeping, sitting, or running), the amount of food energy he takes in, and his metabolic rate.

Heat can be lost by the human body as a result of convective transfers or by radiation to the sky. It can also be lost or gained through conduction by contact with objects that are warmer or colder than he is. Heat can also be gained by convection (from warm air), or can be gained from sources which received radiant energy from the sun and then re-radiate that energy.

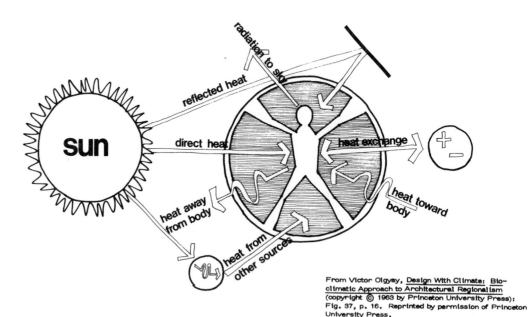

sun

radiation to sky

reflected heat

direct heat

heat exchange

heat away from body

heat toward body

heat from other sources

From Victor Olgyay, Design With Climate: Bio-climatic Approach to Architectural Regionalism (copyright © 1963 by Princeton University Press): Fig. 37, p. 16. Reprinted by permission of Princeton University Press.

Chemical energy
    plants, food, wood (conversion from solar energy
        by photosynthesis)
    man's energy (conversion from other chemical
        energy)
    fossil fuels (long-term conversion and storage
        of radiant energy)
    battery (conversion to and from electrical energy)

Mechanical energy
    internal and external combustion engine (conver-
        sion from chemical energy)

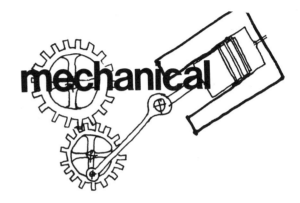

Electrical energy
    electricity (can be readily converted to any of
        the others)
    static electricity (conversion from mechanical
        energy)

Heat energy
    fire (conversion from chemical energy)
    heating coils (conversion from electricity or
        geothermal energy)
    waste heat (by-product of conversion of all forms
        of energy)

There are three ways of transmitting heat energy:
conduction, radiation, and convection.

Conduction is the transmission of energy between
two bodies which are in direct contact. As a common
example, a teapot when placed on an electric stove
receives energy by conduction (or direct contact).

Radiation is the transmission of heat by electromagnetic rays. Only these rays can travel through
a vacuum such as outer space and heat the object
which intercepts them. The sun warms the earth by
radiation which travels in this manner.

Convection is the transmission of heat through a
fluid. An object heats the air (or liquid) in contact
with it, the warmed fluid travels to nearby objects
and warms them. The use of convection for heating
is exemplified by a home forced air furnace. The
furnace warms the air, then the hot air moves to
heat occupants and the house.

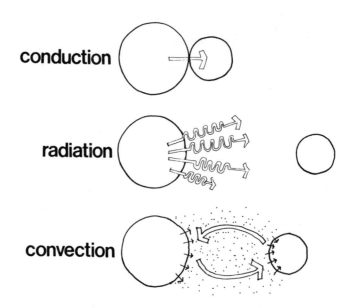

conduction

radiation

convection

energy content

hot

cold

A change in temperature of a body indicates a change in the energy content of that body. This energy is usually in the form of heat or light but can also be in a chemical form. An increase in temperature signals an increase in energy content. A decrease in temperature indicates a decrease in energy content. Some of the more important units for measuring energy are presented below.

Three units which are used for measurement are the Calorie, the Btu (British thermal unit)*, and the Joule. A Calorie is the amount of energy required to raise 1 gram of water 1° C. A Btu is the amount of energy required to raise one pound of water 1° F. The Joule is the unit of energy, hence heat or work, in the Systeme International (metric) system, also known as the S.I.

* The British no longer use this unit, but it is still used in the USA.

1 kilowatt-hr. = 3414 Btu

1 calorie = .003968 Btu

1 Btu = 1055.87 joules

1 calorie to raise 1 gram water 1° C.

1 Btu to raise 1 pound of water 1 degree F.

Heat can be thought of as being sensible heat or latent heat. Sensible heat is the heat a body gives off without changing its state or phase, and is measured as a change in temperature. Latent heat is the heat required by a body to change its state or phase without changing temperature.

A change of state or phase of a substance is a change in its physical condition. A substance may exist in any one of three states:

1. Gaseous (water vapor, steam, or air)
2. Liquid (water, fluid, or mercury)
3. Solid (ice, steel, or wood)

Many substances occur naturally in only one state because of the range of temperatures and pressures which exist on the earth. One example of this is hydrogen which normally occurs in the gaseous state. If the temperature is lowered to 250 F below zero, hydrogen changes from a gas to a liquid. (Throughout this book the degree symbol, "0", is not used. This is consistent with accepted engineering practice.)

Latent heat is required to change a substance from a solid to a liquid or from a liquid to a gas. It is given off when the processes are reversed. Under standard atmospheric conditions water will increase in temperature as heat is added to it. Once a temperature of 212 F is reached, additional heat does not raise the temperature, but provides the impetus for a change of phase. Steam may exist at the same temperature of 212 F at sea level. Per unit weight it contains a great deal more energy than water at this temperature.

 **latent heat**

 **sensible heat**

Water gives off only sensible heat when the temperature range is between boiling and freezing. So if water at one atmosphere of pressure is cooled from 190 F to 50 F, the heat removed or given off is sensible heat. If water is cooled from 35 F to 31 F, the heat removed is a combination of sensible and latent heat, since water freezes at 32 F.

All substances have the ability to store heat. Some can store more heat per unit weight than others; the term for this capacity is specific heat. Temperature is a measure of the kinetic energy of the molecules of a substance and is an indicator of the amount of heat stored in it.

The land, sea, and air share in thermal interaction and each influences the temperature of the others. The temperature of the earth's surface determines the temperature of the atmosphere above it. The atmosphere in turn helps to maintain and equalize the temperature of the earth. A coastal city experiences less extreme seasonal and day/night temperature differences than its inland counterpart. This is due to the relative thermal stability of water compared to air and land. Such stability is a result of the high specific heat of water.

The moon has vast temperature differentials and is an example of a planet without an atmosphere. The bright side of the moon (the side which is in direct solar radiation) heats up to several hundred degrees C; the dark side is hundreds of degrees below zero C.

The atmospheric temperature on the earth varies seasonally and with latitude, but doesn't reach the extremes experienced by the moon. The warmth of the atmosphere varies in proportion to the amount of direct solar radiation (insolation) reaching the earth's surface. The water content or humidity greatly affects the air temperature that is perceived. Latent heat is required to vaporize water and add humidity to the air.

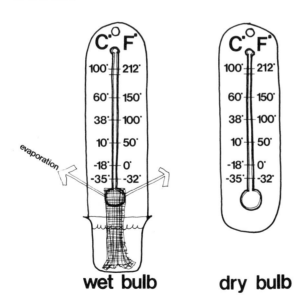

wet bulb   **dry bulb**

The influence of humidity on temperature can be quantified by comparing temperatures measured in two different ways. These are called dry bulb and wet bulb temperature. Dry bulb temperature is a measure which doesn't take into consideration the humidity. This is the method of measuring temperature with which the majority of people are most familiar. The units of temperature are either degrees Celsius, C (often called centigrade), or degrees Fahrenheit, F. The Celsius scale will soon be implemented in the United States as part of a phased program to adopt the S.I. (metric) system of measurements. (The S.I. units will be used in this book, with the more familiar units in parentheses.)

The temperature measured by a wet bulb takes into account the humidity in the air, by measuring the rate of cooling through evaporation at the bulb of the thermometer. When the humidity is low, the wet bulb temperature will differ by many degrees from the dry bulb temperature. As an example, in Arizona during the summer, the dry bulb temperature (the temperature which is announced by a weatherman) may be 42 C (108 F) while the wet bulb temperature may be 21 C (70 F). When the humidity is high the wet bulb temperature will tend to differ only slightly from the dry bulb temperature. St. Louis on a hot muggy day may have a 21 C (70 F) dry bulb temperature and a 20 C (68 F) wet bulb temperature, indicating that the humidity is very high.

The difference between wet bulb and dry bulb temperatures will help explain the use of natural cooling and heating systems in different climates.

The energy relationship between the sun and man's biosphere, and other heat sources outside the biosphere, includes many factors which help to maintain the heat balance of the human organism.

Man can be a receptor of direct or reflected radiation from the sun. He also generates heat from his body processes. The amount of heat he generates depends on his activity level (whether he is sleeping, sitting, or running), the amount of food energy he takes in, and his metabolic rate.

Heat can be lost by the human body as a result of convective transfers or by radiation to the sky. It can also be lost or gained through conduction by contact with objects that are warmer or colder than he is. Heat can also be gained by convection (from warm air), or can be gained from sources which received radiant energy from the sun and then re-radiate that energy.

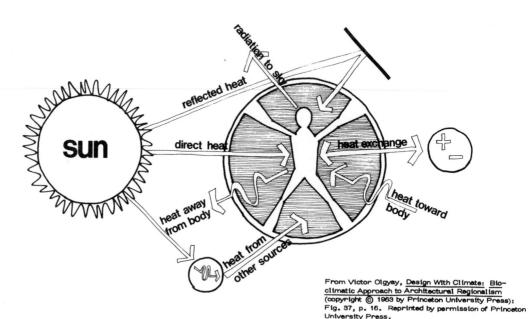

From Victor Olgyay, Design With Climate: Bioclimatic Approach to Architectural Regionalism (copyright © 1963 by Princeton University Press): Fig. 37, p. 16. Reprinted by permission of Princeton University Press.

All of these affect the perceived temperature in a space. Man in different cultures has different ranges of acceptable temperatures for comfort. Many of these are psycho-cultural influences of his society. In the United States the suggested comfort zone is from 20 to 24 C (68 to 75 F), while in England it lies between 17 and 20 C (63 to 68 F). Adaptation of the human to his environment is well illustrated by the contrast between an Alaskan igloo environment where indoor temperatures of 5 to 16 C (41 to 61 F) are tolerated, and a tropical habitat where the comfort zone ranges from temperatures of 22 to 30 C (72 to 86 F).

The comfort zone of tolerable temperatures for the average United States citizen has historically narrowed in range, as a result of increasing dependence upon mechanical systems for comfort. This suggests a psycho-physical conditioning of man to a less tolerable range in proportion to his demonstrated technological ability to control microclimates.

The human comfort zone depends on many factors. A few are listed below:

ambient air temperature: the temperature of the air surrounding the body (the dry bulb temperature)

relative humidity: the percentage of water vapor in the air in relationship to the maximum amount of water vapor it can hold at a given temperature

air movement or speed: how fast the air is moving adjacent to the body

temperature of adjacent objects: the air temperature can be 24 C (75 F), but if the wall and floor temperatures are low, the perceived temperature is less. Sometimes this is called mean radiant temperature of the environment.

ambient air
temperature

relative
humidity

temperature
of adjacent
objects

air speed

air movement

An understanding of the ways in which air moves when subjected to various conditions is necessary for designing efficient energy conserving buildings and systems. The following are simplified air movement principles.

Air will tend to stratify into horizontal thermal layers. As the air is warmed, its molecules will move farther apart decreasing its density. With the decreases in density, the weight of air per unit volume will decrease, causing it to rise above the cooler, more dense air.

air movement increases as the temperature rises

When air is warmed, heat energy is converted to molecular kinetic energy, measured as an increase in temperature. Air pressure differences tend to be equalized by the flow of air from a region of high pressure to one of relatively lower pressure. Wind and weather "fronts" are a result of this equalization.

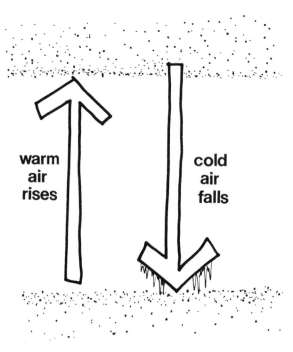

warm air rises

cold air falls

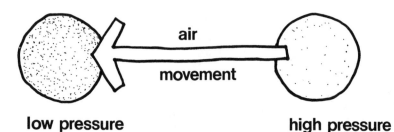

air movement

low pressure

high pressure

This difference in pressure is the basis for many aerodynamic and thermodynamic principles. One of these is the venturi effect, which results from the conversion of pressure to velocity when a fluid flows through a constriction.

Another principle which is useful is that as air passes through a reduced opening, its velocity increases in a "bottleneck" effect and draws adjacent air along with it by induction, upon exiting through the small aperture.

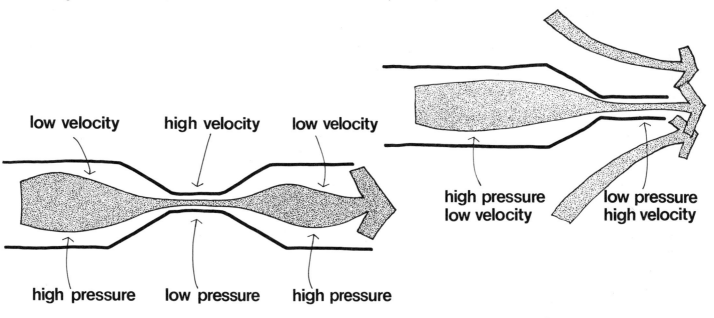

low velocity   high velocity   low velocity

high pressure   low pressure   high pressure

venturi effect

high pressure
low velocity

low pressure
high velocity

light

Electromagnetic radiation which is visible to the eye is called light. Energy from the sun is transmitted in forms other than visible light, which occupies a comparatively narrow band of the electromagnetic spectrum between the short wavelength ultraviolet and the long wavelength infrared. Because the source of the earth's natural light, the sun, is approximately 148,990,000 kilometers (92,600,000 miles) away, the light arriving here is thought of as rays which are parallel to each other. These rays can be manipulated with the aid of lenses to intensify or diffuse the light energy. They can be bent or reflected, using a suitable surface. When light is reflected from a polished surface, the angle at which the light hits this reflecting surface (called the angle of incidence) is equal to the angle at which it rebounds from the surface (called the angle of reflection).

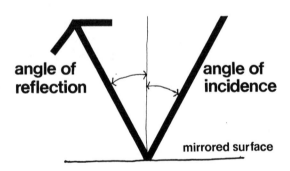

Materials may either absorb, reflect, refract, or transmit light. Colors are a result of certain wavelengths being absorbed and others being reflected or transmitted. To reflect all the colors contained in "white" light a surface should have a coating or color which does not absorb or scatter any of the light rays. White rocks have a surface color which will reflect

light, but because of the irregularities of the surface, they will not reflect the rays in a parallel or predictable manner. Scattering (diffusion) of the light is the result. A mirror, by contrast, produces predictable (specular) reflection, as shown by the adjacent illustration.

If an object absorbs all visible incoming light energy and reflects none, it appears black. If it reflects all the light and absorbs none, it appears white. A leaf reflects green, while absorbing all the other colors, hence appearing green. The visible light spectrum (or range) is made up of different wavelengths, each wavelength producing a different color.

Light and other radiant energy can be controlled and collected. It can be reflected from one surface onto an absorbing surface, thereby intensifying the density of the energy. Parallel rays of radiation can all be focused onto one point or line of points with a parabolic reflector.

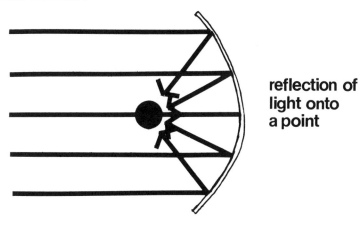

reflection of light onto a point

Radiation can also be bent or refracted with the use of lenses. Lenses can be used for concentrating or diffusing the radiant energy.

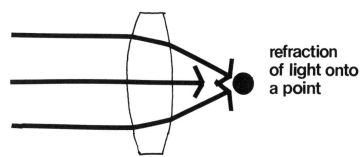

refraction of light onto a point

A thin-section lens used in solar energy collection is the Fresnel. The principle of the Fresnel lens is to duplicate the curvature of a much thicker lens on a flat sheet. The curvature of the large lens is divided into segments and duplicated exactly by the Fresnel lens.

regular lens

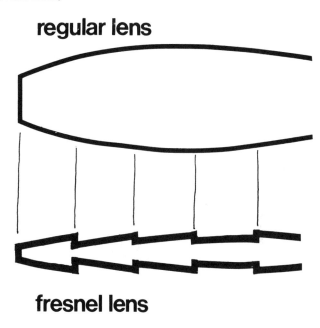

fresnel lens

notes

Natural
Energies

All ecological systems on the earth are inter-related and interdependent. Changes in one system have influence on others. Man has manipulated and changed his natural environment to reflect his own physical comfort zone limitations.

As our belief in technological dependence changes from exploitation of the earth and its biosphere to one of living in harmony with nature and utilizing the natural forms of energy which are continuously available to us, we will use those forms which effect minimum disruption to natural cycles of the earth. With our buildings, we can employ such forces to minimize the pollution and degradation of the environment.

The approach to energy in building advocated throughout this book may be summarized in the analogy of civilization to nature, in which the greatest complexity offers the greatest survival potential. As an example, consider the sophisticated interdependence of all living organisms inhabiting a rain forest or jungle. This is nature at her most intricate and advanced: there are safeguards against all forms of damaging weather, insect blights, and intervention of predators. In fact, there are many duplicate systems which allow such a forest/jungle to survive the extinction of any one species. Nowhere else is the term "balance of nature" more appropriate.

By contrast, the Valley of the Nile River in Egypt shows very elementary life forms and terrain, in which one major storm can threaten an entire year's crop and the human population which depends on it. Civilizations surrounding the Nile have never escaped the cyclic extremes of flooding and drought. No long-term equilibrium exists here.

The analogy to mankind's energy needs is clear: the most complex power source network with the greatest number of back-up systems offers the greatest continuity and survival value to the daily business of any civilization. An oversimplified fuel economy is as vulnerable to changing conditions as a single-crop land area.

In the case of fossil fuel consumption by mankind, increased demand and consequent depletion are particularly serious because these commodities are not renewable. The customary reliance on one fuel source must not simply shift from oil to, say, uranium, or coal, for these are finite also. Such renewable resources as timber are more valuable as building materials than as fuels, and even if diverted only to use as fuels, would fail to keep pace with accelerating (or even present) combustion fuel demands.

A proper concern, therefore, is to optimize a balance of inter-supportive, redundant natural energy systems and energy conservation features in the process of total urban and community planning.

rain forest

The most promising survival community network includes many dependable fuels and materials, similar to the redundant systems of rain forests and jungles.

On a large scale, this means the development of every fuel type known to be safe, with long-range priority given to those which are least finite. Various forms of solar energy and lunar gravitation (creating ocean tides) will remain available for millions, perhaps billions, of years hence.

On a smaller scale, in individual dwellings and other buildings, survival entails the use of as many conservation features as practicable, to allow a smooth transition from one fuel source to another. Thermal efficiency of a building is directly related to implementing the most inter-effective balance of redundant systems.

The health and well being of man depends largely on responsive architecture and the utilization of nature in buildings. The natural energy systems of the earth can be divided into four major areas.

1. Sun   –   energy derived from the sun

2. Wind –   dynamic movement of air caused by temperature differences across the planet's surface

3. Water –   movement and cycling of water influenced by sun and weather

4. Earth –   energy due to natural mass and heat of the planet

Natural energy forms can be used passively and actively. In active systems, motorized mechanical devices are necessary to utilize the energy source, while in passive systems, the need for motors is eliminated by various applications of physical principles, such as the vertical stratification of air temperatures.

A solar system which uses motors to pump fluid through a collector is an example of an active energy collection system. A solar collector which uses the natural thermosiphoned movement of air or water in its heat collection is considered a passive system. Some energy collection and storage systems use both passive and active devices.

desert

sun

The direct energy from the sun has many different architectural applications. Some of these are:

Heating of buildings — Both passive and active solar energy can be used to heat internal and external areas in our built environment. The use of the energy which naturally falls upon a building can bring about less dependence on fossil fuel systems.

Cooling of buildings — Solar cooling of buildings can be achieved mechanically through the use of absorption cooling methods. Evaporative cooling systems depend on water vaporization to lower the temperature of the air, and are most effective in dry climates.

Hot water heating — Collection systems can utilize the sun's heat to supply domestic hot water, to supplement existing hot water systems, or to heat swimming pools.

Ventilation — The use of the sun's energy for inducement of air movement can be an economical method of passively ventilating a building. Depending on the relative elevations of intake and outlet, summer cooling may also be induced.

Ventilation air tempering — Incoming cold air can be tempered (preheated or precooled) before entering the building.

Humidification — Using the sun's energy, water can be evaporated to provide humidification of buildings.

Dehumidification — With the use of rechargeable desiccants, solar energy can dehumidify building spaces.

Desalination and distillation — Through the use of a solar still, salts and minerals can be removed from water.

Drying or dehydration — Food and grains are preserved and lumber is prevented from warping when dried.

Solar responsive architecture — Architecture can use design of buildings themselves to act as efficient solar collectors during cold winter months and as cooling chambers during periods of warm and hot weather.

Phototropic change - Change in color or value through solar radiation could make buildings responsive to their thermal need, such as dark in winter and light in summer.

Landscaping and site planning - Design with nature provides an effective response for all seasons of the year, by coordination between the building and its site. Every tree, bush, and ground cover affects the thermal response of the building.

Natural lighting - The use of natural lighting by the sun can considerably reduce electrical energy usage by buildings. Daylight can be effectively introduced through roofs as well as outer sidewalls of buildings. Reflective external surfaces can be used to increase interior daylighting.

Solar furnaces and ovens - Concentrated solar energy can be used to cook food or provide heat for industrial purposes.

Photo-chemical reactions - Energy from the sun causes chemical reactions in plants, animals, and materials, bringing about changes in color, form, and growth. Greenhouses can effectively extend the growing season for various crops, lessen transportation distance for distribution, and increase food supplies.

Germicidal reactions - The sun's ultraviolet radiation is germicidal by nature. With the proper compass orientation, this can be used to great advantage in certain rooms in all buildings, an obvious example being hospitals. Glass or plastic windows that pass full solar spectral radiation are needed for this process.

Electrical power generation - Use of photovoltaic cells to convert the sun's energy directly to electricity has so far been confined to space vehicles and geographically remote locations on earth. Increasingly broad feasibility is likely, however. Indirect electrical conversion uses heat concentrators in developing steam to generate electricity and is presently more economical for large-scale applications than is photovoltaic conversion.

Switching - Sunlight can be used to activate photo cells controlling switching devices. This application is already in use with lighting for city streets, shopping center parking lots, and building exteriors.

wind /air

The movements of air and wind can be used in many active and passive energy systems. The air moves to equalize any temperature or pressure differentials caused by the sun. The wind can be used for:

Water pumping – Wind driven pumps for wells and irrigation were familiar on the rural American landscape in the early 1900's. Due to expanding rural electrification, windmills were later supplanted by electric pumps.

Grain milling – Many of the old European windmills were used for grain milling. The slow rotary action of the windmill provided high torque for turning of the large milling stones.

Mechanical power – The turning force of a wind machine can provide direct drive power for factories and industries.

Electrical generation – Wind machines can generate electrical power for homes and buildings. Large scale generating plants, using wind machines or vortex generators could provide power for towns or cities.

Wind responsive architecture – The shapes and forms of architecture can be arranged to control external and internal wind-induced ventilation, and to use prevailing wind patterns to power wind turbines.

Combustion air – Oxygen in the air promotes the combustion by fire of various substances or their reduction by heat.

Fresh air – One of the most immediate, continual needs for the sustenance of humans, plants, and oxygen-breathing life forms is fresh air. For these organisms, air pollution has serious long-range health implications.

Cooling media – Air which is cooler than an object in the airstream will carry away heat being given off by the object. The rate at which heat is carried away increases as the velocity of the air increases.

Dehumidification – Air can be used for cooling by first passing it over a desiccant, a moisture removing substance, or by allowing its moisture to condense on a surface.

Humidity – The addition or subtraction of humidity in the air can provide temperature changes necessary for natural heating and cooling. The sensation of comfort is greatly influenced by wind speed and humidity level.

Air support – Structural support by inflatable and floating structures can use the compressive potential of air in confined spaces.

Movement – Wind provides the movement of air to induce the propulsion of ships, gliders, or other vehicles. Pneumatic air can also be employed to move objects, actuate control systems, and propel devices through tubes.

Compression – Energy can be stored in air by means of compression. Containers could be filled with air which is later released to power turbines and generators during peak electrical demand periods.

Insulation – Still air can be used as insulation. To keep the air from moving, pockets or chambers can be designed to minimize air currents thereby minimizing convection heat losses. Closed-cell plastic foams are widely used as insulation because of their ability to entrap air.

Purification – Landscaping, using the texture of trees, bushes, and ground cover, has a purifying effect on air. Plants withdraw carbon dioxide and provide oxygen in the process of photosynthesis. Air can also be purified by fiber, mechanical, and electrostatic filters.

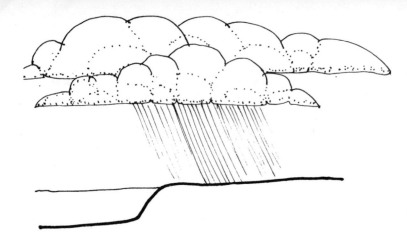

water/ precipitation

The natural cycle of water and precipitation is a force needed to sustain life on earth. About three-fourths of the planet's surface is water. Many forms of energy can be developed from the movement of water in its natural cycle.

Life support - Approximately 2/3 of the human body is water. Water is a vitally needed life sustaining substance for all organisms.

Plant support - The circulation of water through plants and over the land maintains the food supply for the planet, and through photosynthesis, provides our oxygen. Water can be retained in ponds or cisterns for landscaping and greenhouses.

Cooling - Water can be used as a natural cooling medium. As cooler water runs over a warm surface it draws off heat, cooling the surface. Water can be sprayed over surfaces to remove heat by evaporative cooling. Ground water or cold lake water can be used to cool building spaces. The circulation of cold water can be used to reduce air conditioning costs. Ponds on building roofs or on land near buildings can be used for cooling by evaporation.

Humidification - Water can be evaporated to provide humidification of buildings.

Thermal storage - Each gallon of water can hold 660 Btu of heat energy, more than 4 times the heat of an equal volume of 2" gravel. Nevertheless, this property for solar energy water storage is not always a decided advantage due to its thermal diffusion. Gravel storage with equal energy can usually maintain higher concentrations of temperature.

Air purification - Water in the form of precipitation removes pollutants from the air. Rain and snow in some areas are suitable for consumption, while in major cities air pollutants in rain cause impurities which make such water nonpotable. Fountains and waterfalls provide airborne beneficial negative ionization of moisture vapor.

Cleaning - The movement of water acts as a natural cleaning mechanism, dropping heavier impurities as it moves across stream and river beds. Algae and micro-organisms which live in the water can help clean it. Purification of water can also occur by osmosis through a suitable membrane.

Thermal inertia – Large bodies of water such as oceans, lakes, and broad rivers naturally moderate day-to-night temperature changes internally and on adjacent land areas. Coastal and riverfront cities are known for their mild climates and refreshing breezes. Covered swimming pools can be used as heat sinks for solar energy thermal storage.

Recycling – There is a fixed supply of water on and surrounding earth which is continually recycling in one of its three states: liquid, vapor, or ice. More concern with water and how it recycles in the earth's biosphere will help to curtail wasteful uses of the water supply. Waste water can be filtered and re-cycled for buildings.

Electrical generation – The generation of electrical energy is facilitated economically by hydro-electric dams and their turbine generators. It should be noted, however, that dams can have the negative effect of damaging the ecology and of causing over-salinization of irrigated land.

Hydrogen – Water can be reduced to its components of hydrogen and oxygen by electrolytic action. Hydrogen can be stored or directly used as a clean, gaseous, powerful fuel. Oxygen can be similarly used as an aid to the combustion of other substances. Hydrogen fuel cells are an advanced technology needed for energy conversion.

Mechanical power generation – Water wheels can be used for rotary power generation. The force from turning water wheels was used in the 1700's and 1800's to run weaving and other equipment. Also, the force derived from wheels was used in early industrial plants and for grinding ore in mines.

Ocean thermal gradients – Because the oceans cover approximately 75% of the planet's surface, they are the largest receptors of solar energy. The thermal gradients of the ocean (temperature decreases with depth) can be used to power heat engines which could produce great quantities of electrical power.

Tidal power – Electrical power can be produced by harnessing the movement of tides with dams and generators. Tidal power stations are most efficient when using reversible turbines which take advantage of the opposite directions of movement of ebb and flow. The gravitational force of the moon causes tides, and this type of application is an example of "lunar energy". Very few sites in the contiguous United States are suitable for electrical generation using tidal power. Worldwide, however, a number of locations appear worthwhile.

earth

The earth can be thought of as an energy source, but its quantity of energy is very limited in comparison to that of the sun. Some forms of energy from the earth and manners in which the earth may be beneficially used are:

Life support – Countless organisms and micro-organisms find a home within the earth. Nutrients in the soil, formed and released by nature's regenerative processes, nurture concentrated life forces. The habitat constructed in response to the natural environment by these creatures offers a remarkable example for the design of homes and buildings.

Resource container – The earth provides minerals and materials which can be used as energy sources. Fossil fuels are contained and formed under the earth. These are renewable only on an extremely protracted time scale, measured in millions of years.

Filtering – Sands and gravel on the earth act as a filter by purifying water. Many sewage treatment plants use varying sizes of stone and gravel as filter systems.

Purification – As water percolates down through the soil, impurities are removed. The process is aided by aerobic and anaerobic bacteria that reduce wastes to a harmless state.

Geothermal – The earth is a natural heat generator producing intense heat at its core, which is conducted out to the surface and into subterranean water.

Gravity – The force of attraction between two masses can be used to store energy. Off-peak electrical power, as an example, can be stored as potential energy by pumping water to a high level retaining area (water tank or dam). The gravitational force acting on the water after it is released will convert the potential energy to kinetic energy, which will ultimately power a turbine for the generation of electricity.

Interior thermal mass – In buildings, large interior masses of earth, concrete, or masonry which are insulated from ground temperatures and from changes in external climatic conditions will stabilize internal building temperatures.

Thermal inertia – The planet as a whole exhibits strong thermal stability from the ground level downward. In most temperate climatic zones, a "frostline" is found at about one meter depth, below which the soil never freezes.

Solar heat absorption – The earth, in varying degrees, is a natural absorber of solar energy. Although water accounts for 75% of its surface and retains solar heat effectively, certain land masses are more efficient absorbers. This is particularly true in equatorial regions, where the earth's surface is most nearly perpendicular to incident radiation. Rain forests and jungle growth also trap heat.

Building material – Stabilized earth, adobe, rammed earth, earth floors, sod roofs, and mud surfacing are useful as low cost building materials. The energy expended for manufacturing and transportation are reduced by using excavated on-site materials for such structures.

Insulation – Earth can act as an insulator by tempering heat losses and heat gains through roofs, walls, and floors. The earth can be manipulated into forms to protect structures from wind and other harsh weather conditions. A mass of earth can also act to protect plumbing pipes from freezing and reduce excessive heat losses from building ducts.

Earth concrete – Dry cement mixed with earth and sprinkled with water can provide a durable paving surface.

Thermal cooling – At depths of 4' or more earth temperatures usually remain at about 50 F to 60 F. During warm and hot weather this can be used as an effective medium for cooling buildings.

Decomposition – The waste products of living earth organisms and from their activities are decomposed into combustible gases such as methane which can be used as an energy source.

notes

Sun

A star in the universe, the sun is a giant nuclear fusion reactor. At a speed of 300,000 km per second (186,000 miles per second), the energy from the sun takes 8.3 minutes to reach the earth. The sun is extremely large, with a diameter of 1,390,000 km (865,000 miles) and a volume a million times greater than that of the earth. The temperature of its photosphere or surface is approximately 5760 C (10,400 F). Nuclear transformations at the core produce intense heat and consume about 3.6 billion kilograms (4 million tons) of mass per second. The sun acts as a fusion reactor combining 4 nuclei of hydrogen into one. This results in a decrease in mass. There is no need for alarm because the supply of hydrogen will probably last another 6 billion years.

The sun's core radiates heat as a result of its nuclear reactions. This is transmitted by radiation in areas closest to the core; heat is then transmitted by convection up to outer layers, where it is radiated outward. Since the sun is extremely far from the earth (150 million kilometers or 93 million miles), and the earth is small in comparison to the sun, solar radiation is considered as traveling to earth in parallel rays.

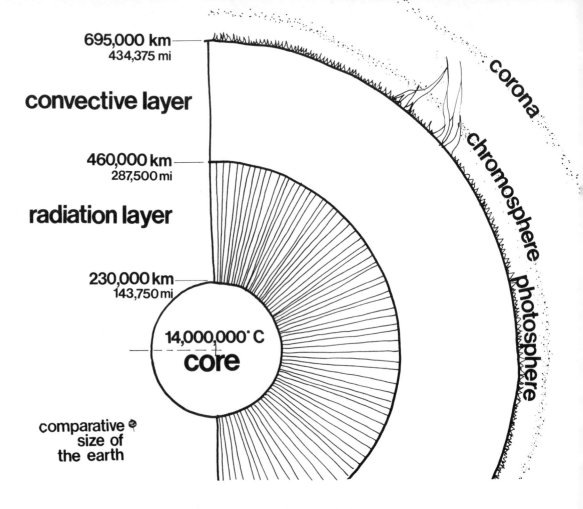

convective layer

695,000 km
434,375 mi

460,000 km
287,500 mi

radiation layer

230,000 km
143,750 mi

14,000,000° C
core

comparative
size of
the earth

corona

chromosphere

photosphere

cross-section of the sun

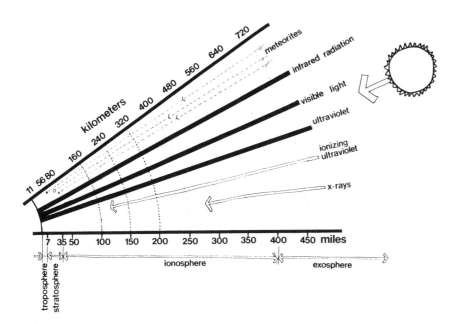

Radiation from the sun has catalyzed or directly supplied most of the natural energy systems on the earth.
Many different levels of energy from the sun continuously flow toward the earth. Not all of these different
radiative frequencies penetrate the atmosphere and
reach the earth's surface. This is because the atmosphere is composed of many layers, each with its own
composition of oxygen, nitrogen, hydrogen, and other
matter. These layers protect the earth and help sustain the life forms on its surface. Visible light makes
up a portion of the sun's energy which is able to penetrate all the different levels of the atmosphere,
whereas much of the ultraviolet and infrared energy
(which we cannot see) is absorbed or reflected by it.
The upper levels of the atmosphere contain ozone
which absorbs dangerous ultraviolet radiation and
x-rays. Most meteors and other debris from space
burn up, never striking the earth, because the atmosphere is more dense near the surface than it is at
higher altitudes.

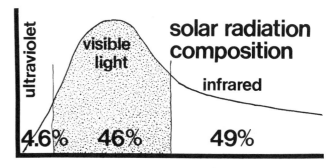

43

Some of the solar radiation which pierces the atmosphere is reflected by the earth's surface or by cloud cover. Some of it is absorbed by the lower levels of the atmosphere, causing warming of the air. This sets up a temperature differential which generates air movement or wind. Major currents or patterns of the wind are a result of the earth's rotation about its axis, and day-night cycles of solar heating and subsequent cooling.

A great amount of energy is absorbed by the earth's surface itself, as well as by oceans, plants, and buildings. The energy which plants absorb is potentially convertible into food energy, material for combustion (logs, etc.), or fossil fuel (long-term decay). The energy absorbed by the oceans causes water evaporation and motivates the complete water (hydrologic) conversion cycle.

In combination with the earth's motion around the sun, the water cycle and the wind provide weather and the change of seasons.

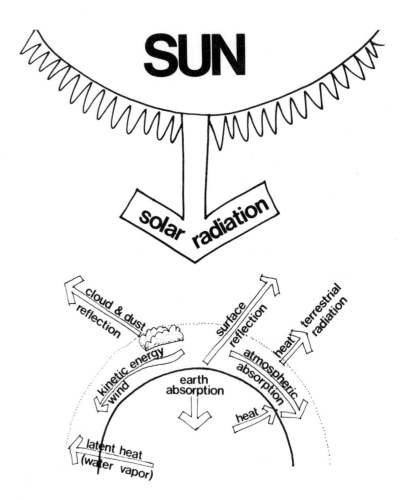

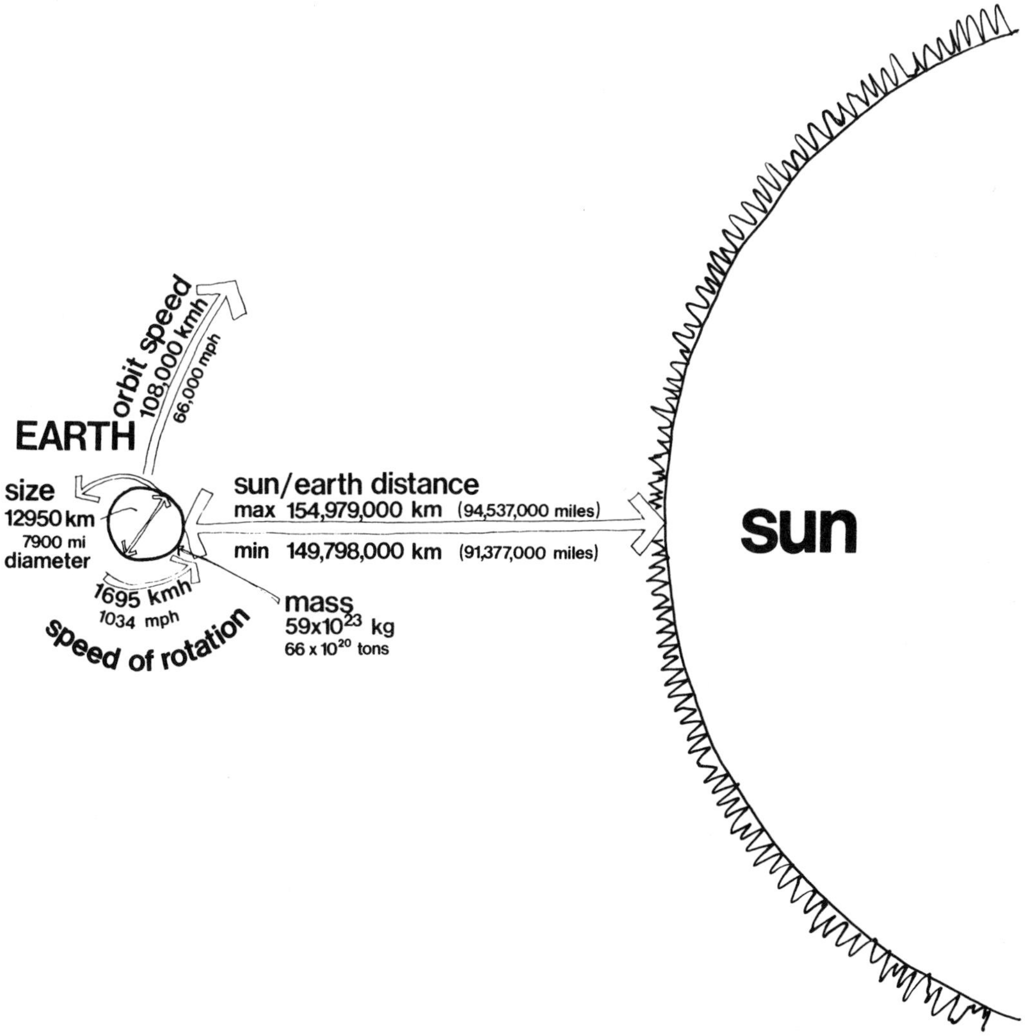

EARTH

orbit speed
108,000 kmh
66,000 mph

size
12950 km
7900 mi
diameter

sun/earth distance
max 154,979,000 km (94,537,000 miles)
min 149,798,000 km (91,377,000 miles)

1695 kmh
1034 mph
speed of rotation

mass
59x10^{23} kg
66 x 10^{20} tons

sun

Some of the physical characteristics of the earth and the sun include their distance apart, their masses, the earth's speed of rotation, and its orbital velocity. The earth is not a static, stable object, but a moving, dynamic body protected by only a thin atmosphere. It's a spaceship planet, traveling through infinite void, powered by the sun's energy which provides it with food, water, oxygen, and fuel.

The earth is a giant spacecraft, supporting life as it orbits the sun. The sun provides the earth with the energy it needs to sustain and maintain life. The earth with its thin atmosphere (a layer only 160 km or 100 miles deep) supports almost four billion people. Smaller spacecraft in many ways duplicate the conditions and functions of the earth, utilizing the sun's energy and shielding their occupants as they travel through space. Unable to shield themselves with atmosphere as does earth, however, these craft must rely on sophisticated metallic alloy skins and sensitive thermal conditioning systems. Many use photocells to collect the sun's energy and convert it to electricity. The energy received on the earth's surface in one day exceeds the human consumption of energy in that period of time by a thousandfold, and will continue to do so throughout the foreseeable future.

Both the position of the earth and locations on the earth greatly affect the feasibility of using natural energy systems powered by the sun. The earth rotates and in so doing appears to cause the sun to "rise" and "set". The angle of the sun and the intensity of its radiation are affected by our location on the earth. The earth's axis of rotation is tilted at 23½° from perpendicular to the plane of its orbit. Geographic latitude (the angular distance measured from the equator) affects the angle at which solar rays strike an object.

The earth also travels around the sun, completing one orbit every 365.24 days. The tilt of the earth in combination with its orbital motion around the sun brings about seasonal changes in the northern and southern hemispheres.

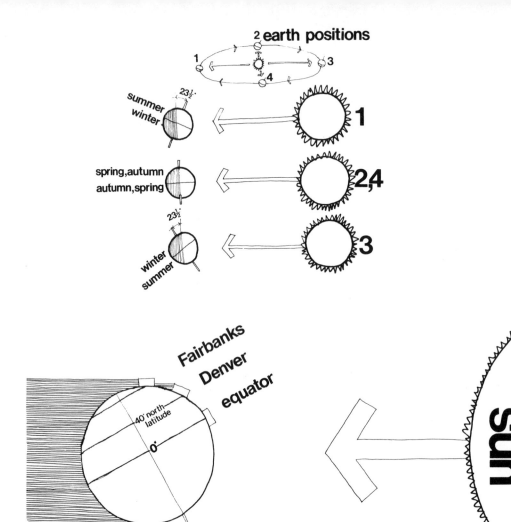

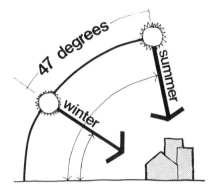

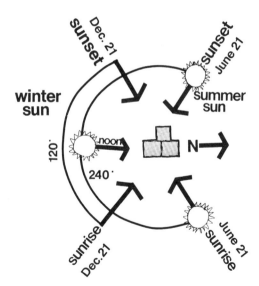

typical for 40° north latitude

The apparent path of the sun changes every day. For example, at the winter solstice position (December 21 is the shortest or solstice day) at 40° north latitude, the sun travels through only a 120° azimuth or plan arc, from dawn to dusk. On June 21, the summer solstice day, the sun's azimuth arc totals 240°. Occurring between solstices are spring and autumnal equinoxes, in which daytime elapsed time equals nighttime elapsed time. The sun appears to travel on a curved path from sunrise to sunset, with the zenith occurring at noon. The difference between the maximum solar altitude in the summer and the minimum solar altitude in the winter is 47°.

Orientation with respect to the sun affects the amount of insolation a surface will receive. Considerations when designing solar heat collection systems are: the tilt angle of a fixed solar collector, the plan axis on which the building is oriented, and any opaque obstructions which interfere with solar radiation, especially during the hours from 9:00 a.m. to 3:00 p.m.

The tilt angle for maximum solar collection, by a collector in a fixed position, depends on the time of the year when the greatest amount of the sun's energy is needed. For maximum winter collection (for heating purposes) this angle should equal the geographic latitude plus about 15 degrees. For maximum summer collection (to power air conditioning or ventilation equipment) the angle should equal the latitude minus about 10 degrees, while a compromise angle for both summer and winter lies between these two angles, the ideal being dependent on climatic conditions. Variations from these figures of up to 20° can be allowed with only a minor loss in collector output.

collecting the sun's energy

A solar collector is a device used to absorb heat which comes from the sun in the form of radiation, and transfer it to a circulating fluid (air or liquid).

The flat plate solar collector consists of a black metal plate covered with glass or plastic and backed with insulation. The metal plate is known as an absorber and may have tubing built in to contain the circulating fluid. An air space separates the cover from the absorber.

The glass or plastic is transparent to incoming solar radiation, yet glass in particular is opaque to radiation of longer wavelengths. Solar radiation passes through the cover and is absorbed by the black surface, increasing the temperature of the metal. Longer wavelength infrared radiation is emitted from the absorber, but most of it cannot pass back through the glass. This produces what is known as the greenhouse effect. (It commonly occurs in automobiles on a hot summer afternoon.)

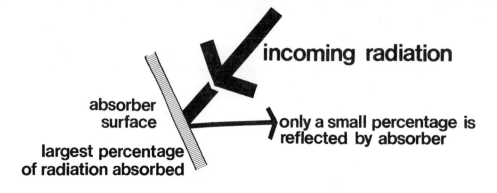

incoming radiation

absorber surface

only a small percentage is reflected by absorber

largest percentage of radiation absorbed

A liquid is circulated through the tubes in the absorber or air is blown behind it. If these fluids are at a lower temperature than the absorber, heat will be transferred to the fluid. Insulation decreases heat loss from the back of the collector.

If the maximum potential of the sun's energy is to be utilized, as much heat energy as possible must be absorbed and retained. To illustrate the basic workings of a flat plate collector it is necessary to examine the way in which energy is transported into and out of the collector.

A typical flat plate collector is an absorbing unit. It cannot yield more energy output than the maximum solar radiation striking its surface.

The glass or plastic cover of a solar collector can either reflect, absorb, refract, diffuse, or transmit incoming solar radiation. If the glazing material is clear, only a small percentage of the incoming radiation is reflected or absorbed while most is transmitted through to the absorber surface. As the angle of incidence increases, a greater percentage of the radiation is reflected from the surface of

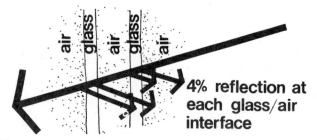

air glass air glass air

4% reflection at each glass/air interface

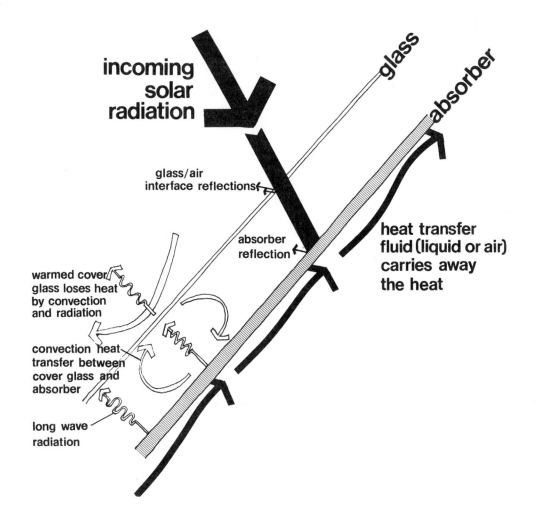

incoming solar radiation

glass

absorber

glass/air interface reflections

absorber reflection

heat transfer fluid (liquid or air) carries away the heat

warmed cover glass loses heat by convection and radiation

convection heat transfer between cover glass and absorber

long wave radiation

the glazing. Boundaries created by media with different optical densities reflect radiation. At each interface between glass and air approximately 4% of the radiant energy is reflected.

Most of the incident radiation remains at the absorber surface while a small fraction is reflected. Glass is much more opaque to long wavelength heat than the plastics available today. Hence, a plastic-encased collector will lose considerably more heat than the same one protected by glass.

It is desirable for the absorber surface to collect as much energy as possible. Since no light is reflected by the absorber surface, it appears black. Plants reject the color green as a means of tempering the intense heat they receive from the sun. In this way they receive only energy within their most beneficial spectral range.

Selective surfaces have been developed to absorb as much of the incoming solar heat radiation as possible (90%) and emit the least possible amount of long wave-length radiation (10%). Some paints offer as much as 99% absorptance. However, these paints and other nonselective coatings allow greater heat losses by reradiation.

As a collector warms up it transfers heat by both radiation and convection. It is desirable to transfer the maximum amount of heat possible to the collector fluid and reradiate or emit as little as possible to the outer surfaces of the collector. The fluid (water or air) behind the absorber plate absorbs heat and then transports it into the heating system. Some of the heat is carried off the absorber surface by convection, and eventually warms the inner surface of the cover glass. It is then conducted through the glass to the outer surface and is lost to the atmosphere by radiation and convection.

Collector temperatures are regulated by controlling the flow rate of their heat transfer fluid (air, water, or others). The faster the fluid flows through the absorber, the cooler its surfaces stay. The heat losses (both by radiation and convection) in proportion to the heat obtained for utilization when operating at very high temperatures are much greater than at low temperatures. Because of these losses, flat plate collectors should operate at low rather than high temperatures. Generally, the range between 40 C (104 F) and 94 C (201 F) is best. Parabolic and other concentrating collectors, by contrast, operate at much higher temperatures (sometimes as high as 1093 C (2000 F). However, they are designed to operate at these high temperatures and have much less absorber surface area than do flat plate collectors of equal projected area.

One thing to consider when designing a collector is the number of glazing sheets to use. Some units have one, but most units have two panes of glass. Some are insulating glass. Water white glass with low iron content allows more solar radiation to pass through it. The number of sheets of glass or plastic used affects the amount of incoming radiation reaching the absorber surface, as well as the amount of

heat lost to the exterior. As the thickness and the number of sheets of glass are increased, the amount of solar radiation which is able to penetrate this barrier is reduced. Certain configurations work best in certain climates. For example, the decrease in heat gain through two panes of glass might have to be weighed against the heat loss increase through only one pane of glass. The most critical climatic feature to consider when making these decisions is the relationship between the ambient outdoor air temperatures and the amount of insolation available. Problems can develop when plastic is used for glazing the collector. Plastic is susceptible to attack by ultraviolet radiation, which may deteriorate the plastic and turn it yellow, decreasing its efficiency. Some plastics cannot withstand the high temperatures that collectors can attain when the flow of cooling fluid is interrupted. Those which can, suffer from extremely high coefficients of thermal expansion compared to glass. This factor in turn modifies collector frame design requirements.

Other considerations in the design include the spacing between the glass plates, the spacing between the absorber and the glass, the thickness of backing insulation, placement of condensation barriers, and the control of collector edge heat loss.

Illustrated are some examples of general types of flat plate collectors, tube collectors, and parabolic collectors. Also included are examples of collection systems and solar powered cooling cycles.

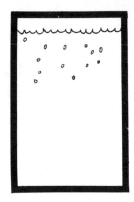

liquid energy storage

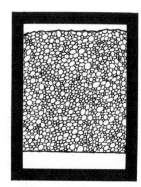

rock bed energy storage

liquid type flat plate collector

Some flat plate collectors use a liquid as the heat absorbing fluid. Water and other liquids have been used for this purpose. Some of the considerations in choosing a fluid for a liquid flat plate collector are:

1. Are the materials of which the collector surface and piping are made susceptible to fluid corrosion?

2. What is the quantity of fluid required and what is its cost?

3. At what temperatures does the fluid freeze and boil?

4. Is there homogeneity among all the materials with which the fluid comes in contact? Electrolysis will occur if two different metals are bridged by a liquid containing acid or dissolved inorganic salts. This will cause chemical corrosion.

Liquid flat plate collectors may have a variety of collector tube configurations. Some of these configurations are:

The liquid is usually pumped into the low side or bottom of the collector. This is done to distribute the water evenly in the system and decrease the trapped air. The sun heats the collector surface. As water travels through the tubes, it picks up heat from the surface by conduction. Heated water then leaves the collector at the top and is pumped to storage or is used in the heating system.

Fluid heat storage volumes tend toward thermal equilibrium, equal temperatures throughout, as a consequence of their natural convection currents, which cause mixing. Temperatures in a tank full of fluid cannot be changed as rapidly as the temperature of one layer of rocks in a rock bin.

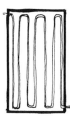
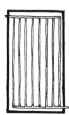

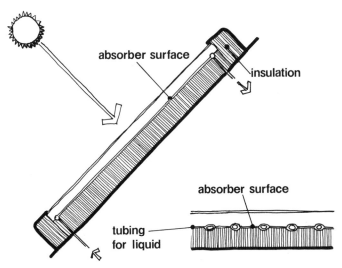

absorber surface

insulation

absorber surface

tubing for liquid

The Thomason "trickle" type collector uses water flowing in the troughs of a corrugated surface to collect heat. The surface is painted black to absorb the maximum amount of the incident radiant energy. Water is supplied through a tube located at the top of the collector, then runs down the corrugated surface (aluminum or galvanized steel). As the water flows over the collector surface it picks up heat. The warmed water is collected by a gutter and is directed into a hot water storage tank.

The air type flat plate collector works similarly to the liquid type flat plate collector. Air is pumped or blown by means of a circulating fan through the manifold for distribution into the collector. The collector surface gains heat from the sun and raises the temperature of the air in contact with it. As the air heats, it rises in the collector. The air exits at the top of the collector, normally at a peak temperature of 54 C (130 F) to 66 C (151 F), and then is supplied to the house heating system or is pumped to the storage bin where its heat is transferred to rocks.

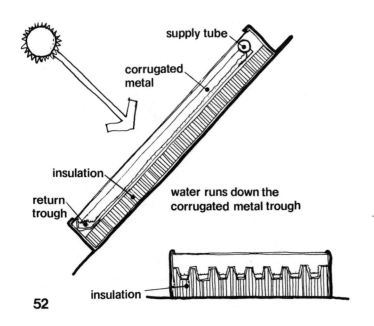

supply tube

corrugated metal

insulation

return trough

water runs down the corrugated metal trough

insulation

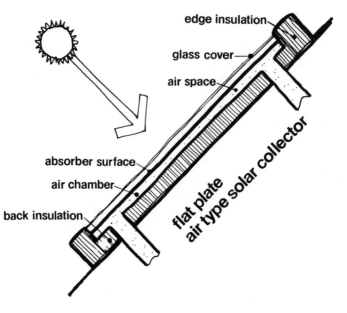

edge insulation

glass cover

air space

absorber surface

air chamber

back insulation

flat plate air type solar collector

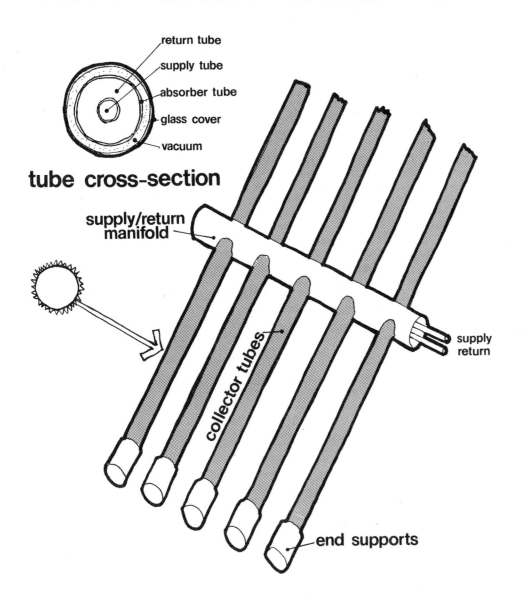

return tube

supply tube

absorber tube

glass cover

vacuum

tube cross-section

supply/return manifold

collector tubes

supply
return

end supports

evacuated tube collector

One type of evacuated tube collector is a recent development which utilizes a series of tubes with special absorptive coatings to collect solar energy. The tubes have the advantage that almost the same amount of surface area is exposed and perpendicular to the sun at any time during the day.

The evacuated tube collector is composed of a series of three concentric cylindrical tubes, with a vacuum (nearly zero air pressure) between the outer and middle tubes and a black selective coating on the outer surface of the middle tube. Water or air (usually water because of its higher specific heat capacity) is circulated from supply pipes through the inner tube. As it travels through this space it picks up heat. Upon reaching one end of a collector module, it enters the volume between the inner and middle tubes where it reverses flow direction and continues to build up heat content. It is then drawn off by the return tube and circulated into the heating system.

An evacuated tube system has the advantages of higher collection efficiency at standard operating temperature and the utilization of high temperatures in its collection process without excessive heat loss. The vacuum between the outer tubes of glass helps attenuate conductive and convective heat loss, but can do nothing for radiation heat loss. This latter loss can be decreased by applying a "selective surface" interior coating to the outer tube.

parabolic collectors

The paraboloidal collector concentrates large amounts of solar energy on a small area. This concentration allows high temperatures to be attained. For effective collection, the parabolic concentrator must track the sun so the sun's rays are perpendicular to the frontal plane of the paraboloid. In the altazimuth system, the unit moves vertically to match the sun's altitude, and tracks horizontally to follow the solar azimuth during the day. Since the sun is in different positions every day, the tracking mechanism must be very sensitive and exact. This is likely to limit such devices to use in large buildings, where capital costs for mechanical systems may be offset by large fuel savings.

Another system to concentrate the solar radiation has a parabolic trough oriented along an east-west axis. The angle of the trough with respect to the horizontal is adjusted weekly to accommodate changes in solar altitude. The reflective trough concentrates energy on an absorber tube through which a heat absorbing fluid is circulated.

The paraboloidal surface of the collector is mirrored and reflective. Radiant energy is reflected onto an absorber area which is usually glass covered to minimize heat loss by convection and radiation. The absorber has a liquid circulating inside it to carry away the heat energy. It is essential to keep reflective surfaces very clean to avoid reflective loss.

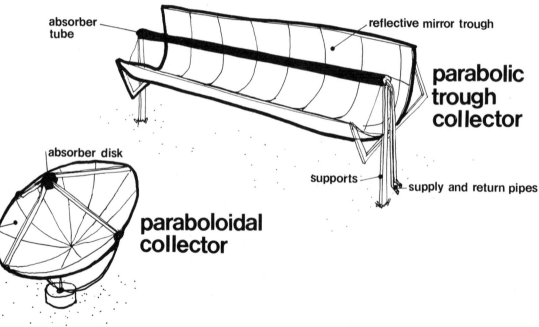

absorber tube

reflective mirror trough

parabolic trough collector

absorber disk

mirrored surface

paraboloidal collector

supports

supply and return pipes

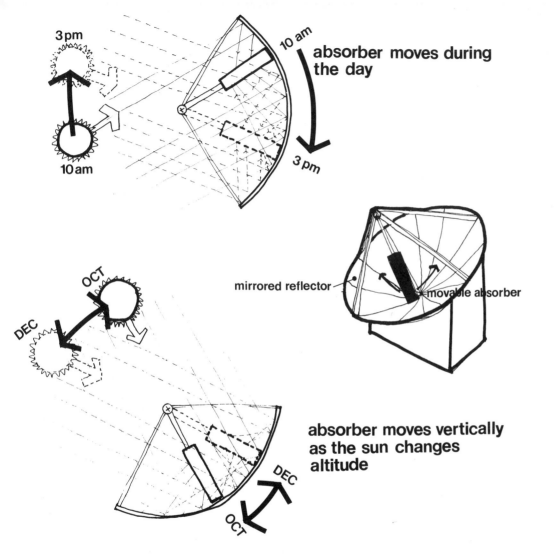

absorber moves during the day

3pm

10 am

10am

3 pm

OCT

DEC

mirrored reflector

movable absorber

absorber moves vertically as the sun changes altitude

DEC

OCT

SRTA collector

The major drawbacks to the parabolic collector are the high costs of optically precise surfaces, the need for expensive tracking mechanisms, and the difficulty in maintaining a clean optical surface. The SRTA (stationary reflector tracking absorber) collector is being developed to eliminate the need for tracking the sun with the mirror reflector. With this collector the hemispherical reflector is stationary and the absorber tracks the sun. The light energy is reflected off the mirror and onto the absorber.

The SRTA absorber is made up of tubing covered with glass. Fluid circulates in the tubing to transfer the heat. Since the temperatures in concentrating collectors far exceed those in flat plate collectors, the fluid used is usually not water. Whatever the fluid, it must nearly always be confined under high pressure to prevent boiling.

air type collector system

A flat plate air type collector system can operate in several modes, some of which may be in operation simultaneously.

1. Heating with the solar collector

    Air in the return duct from a room is routed to the base of the collector. It is then passed through a manifold to distribute the air evenly in the collector. The air is warmed by the sun in the flat plate collector (see preceding section on collectors), then delivered directly to supply ducts and distributed in rooms. When any room has attained its desired temperature, excess heat energy is transported to the storage system.

2. Charging the heat storage

    The heat captured by a flat plate air collector system is stored in a rock bin. The rock bin is sealed in an insulated container which has inlet and outlet connections. When the rooms of a building are sufficiently warm and there is still adequate sunshine for the collectors to operate, heat can be stored for use at a later time.

    To increase the amount of heat stored in the rocks, air is drawn out of the rock bed, routed through the collector and then returned to the rock bed. Cold air leaves the rock bed near the bottom and hot air enters near the top. As warm air enters the top of the rock bed, it

transfers or loses its heat to the rocks. The warm air flows down through the rock bed until it is drawn out and returned to the collector. The rocks will gain heat each time newly warmed air is passed over them. Because each rock in the bin has a very large surface area (smooth, graded pebbles of 1" to 2" diameter are used), they are capable of removing heat from flowing air at an extremely fast rate. Air entering such a bin at a temperature 20° C (68° F) above that of the bin, and at low velocity (less than a meter per second), will lose most of its temperature difference and reach thermal equilibrium within the first 30 cm (1 foot) thick layer of rock. This means that full room heating temperature may be obtained from a rock bin which is depleted to its last 30 cm thick rock layer. The storage process continues until each horizontal 30 cm "layer" of rock has reached the incoming air temperature, and the bin is fully charged. When the sun has set or has been obscured by clouds, and the house requires heat, air is drawn from the hottest air layers at the top of the rock bin.

By contrast with the rock heat storage system, the entire volume of liquid in a liquid heat storage system must be maintained near its use temperature for it to be effective. Such performance differences plus the difference between the sizes of ducts and pipes and the power needed by fans and pumps would lead to a preference for heat storage in a solid medium rather than a liquid, for most building applications involving solar collection in hot air. Moreover, air-to-liquid heat exchange is less efficient than air-to-air, liquid-to-air, or liquid-to-liquid.

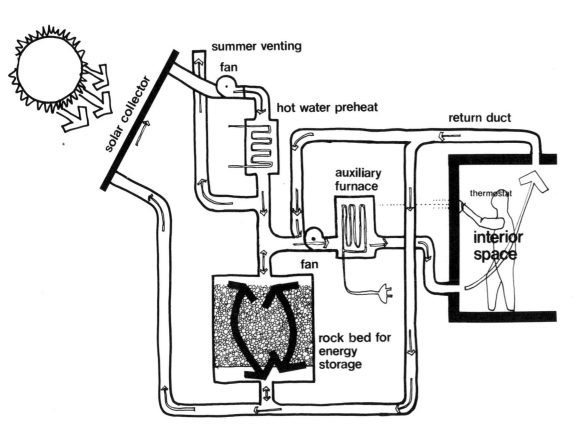

air type collector system

3. Heating with rock storage

   Air is drawn out of a room via the return duct, then routed to the storage bin (which was previously solar heated), increases in temperature as it passes through the rocks, and is then supplied back to the room. As the air is drawn through the rock bed (from bottom to top since heat will stratify in rocks), it gains the heat which the rocks have stored. This heat is then supplied to the room.

4. Hot water preheat

   In most climates, domestic hot water for homes and other buildings can be preheated using solar energy, year round. Warmed air from the collector can be drawn across a heat exchanger to warm the water to a temperature requiring little or no fuel to reach final use temperature. The warmer the air from the solar collector, the higher the preheated water temperature. In the summer, when solar collectors are unnecessary for room heating, they may be fully applied to domestic water heating, and in most cases will raise water to final use temperature.

57

5. Auxiliary heating

An auxiliary heating system is required during times when the solar system cannot supply all the heating requirements of a building. This can be during periods of limited sunshine or abnormally cold weather. If the supply of air to a room from the collector or storage system is not warm enough, an auxiliary heating system should automatically activate. Most storage systems are designed to meet a normal mid-winter demand for about half a day (essentially overnight).

6. Summer venting

In the summer when warm air from the collector is not needed to heat a building, the system should be vented. This is usually done automatically. The warm air near the ceiling of a room is drawn out and supplied to the solar collector where it is heated further. It can pass over the hot water preheat heat exchanger, and be routed out of the building. The warm air which is drawn from the building is replaced by cool air from the outside through dampered vents.

liquid type collector system

A flat plate liquid type collector system can work in several different modes.

1. Collecting heat

A liquid, antifreeze solution or water, is pumped to the base of the collector (see section on collectors), from which it is distributed uniformly behind the absorber plate. The temperature of the liquid increases as it takes on heat from the absorber.

2. Storing the heat

Depending on the type of system, one of two routes is taken by the heated liquid. If the liquid is an antifreeze solution, it travels to a heat exchanger where its heat is delivered to the water in the storage tank. If the liquid is water, it is routed directly to the storage tank, where it mingles with the heated water already there. Both these methods raise the temperature of the entire storage tank, since convective currents in the tank will tend to bring the entire tank into thermal equilibrium.

3. Using storage for direct room heating

When necessary, the building may be heated using only the heat in the storage tank. Water is pumped out of the tank and circulated through a fan-coil unit or other heating device where it loses much of its heat to a room. The cooled water is then routed back to the storage tank.

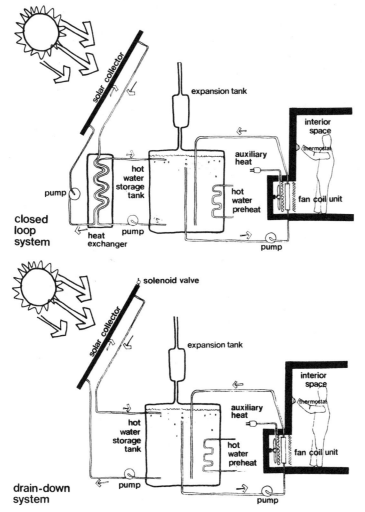

closed loop system

solar collector

expansion tank

interior space

thermostat

auxiliary heat

hot water storage tank

hot water preheat

fan coil unit

pump

heat exchanger

pump

pump

drain-down system

solar collector

solenoid valve

expansion tank

interior space

thermostat

auxiliary heat

hot water storage tank

hot water preheat

fan coil unit

pump

pump

liquid type collector systems

4. Auxiliary heat

When the temperature of the collector (due to limited sun conditions) or the temperature of the storage is too low for space heating, the system should automatically activate an auxiliary heat supply. Depending on the type of system, this may heat the water before it enters the fan-coil unit, or may be a forced air system.

It is necessary to have an "expansion" tank located near the heat storage tank to protect the main tank from any damage that could be caused as a consequence of hot water having a higher specific volume than cold water.

5. Preheating domestic water

By using a heat exchanger, the stored heat can be used to preheat domestic water. In the summer, this system can supply almost 100% of the domestic hot water. Another excellent use of this energy when it is not required for space heating is to heat swimming pools.

photovoltaic collector system

The photovoltaic collector system converts radiant energy directly into electricity. Photovoltaics have been developed mainly for space application where size, weight, and independence of all systems are more critical than their cost. For all other applications, they have the disadvantage of being extremely expensive. New techniques for growing the crystals needed for photovoltaic cells should decrease their cost, as should fully automated assembly systems.

In some photovoltaic systems the photocells are mounted on a surface with a water type solar collector behind them. The water or fluid will help to keep down the temperatures of the cells, while functioning as a conventional heat collection medium. The output of these cells increases as temperature rises to about 82 C (180 F) and then drops rapidly.

Photovoltaic cells will generate DC (direct current) electricity (the electricity in homes is AC - alternating current) which can be stored in batteries or fuel cells. Wet cell batteries store electricity as chemical potential energy by charging particles and plates in a solution; they can be recharged many times, but most have a life expectancy of about three years. Photovoltaic systems can operate at different voltage levels, and may provide AC power through batteries using inverters.

Silicon for photovoltaic cells is "grown" in large single crystals, then wafer thin strips are carefully sliced from the crystal to form the basic unit of the cell. The silicon wafer is then coated with another material, such as boron, to produce a positive electrical layer which interacts with the negatively

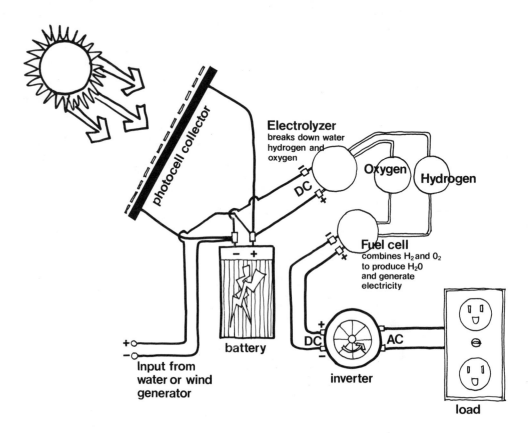

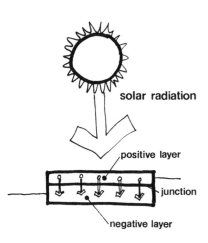

solar radiation

positive layer

junction

negative layer

photocell

charged silicon layer. The interface between the two layers forms the p-n (positive-negative) junction.

When photons, the basic energy particles of light, strike the silicon cell, they are converted to electrons in the p-n junction. The positive layer accepts the negatively charged electrons, and the negative layer rejects them, producing direct current (DC) electricity.

In fuel cells hydrogen and oxygen are combined to form water and electricity. In the opposite process, electrolysis, water is separated into hydrogen and oxygen, using an electric current. This results in energy storage in the hydrogen and oxygen components. Fuel cells produce direct current (DC) which can be converted to alternating current (AC) by an inverter.

An auxiliary generating system such as a windmill or watermill can be added to a photovoltaic or fuel cell system since they both generate electricity.

eutectic salts

Latent heat is absorbed or released when a substance changes phase or state. This was explained in a previous chapter. This is the principle that is exploited when eutectic salts are used for energy storage. Compared to most other materials, they will store a large amount of heat in a relatively small volume. The melting and boiling temperatures of these salts are separated by a wide range. Once they have melted they can reach very high temperatures before vaporizing.

The salt starts out as a solid. As heat is added to it, its temperature rises until it reaches the temperature at which additional heat causes it to become a liquid. A large amount of heat is stored in this transition. When heat is removed from the liquid salt it will solidify, giving off the latent heat which was stored in it.

Heat from a solar collector can be stored as latent (phase change) heat in eutectic salts. When heat is required for space heating or other purposes, and it is not available directly from the collector, it is retrieved from storage causing the salt to reform as a solid. Theoretically, the salt can cycle from a solid to a liquid and back an infinite number of times, taking on or giving off latent heat at each phase change.

There are many problems encountered with eutectic salts which have kept them from gaining widespread acceptance as a heat storage material. Many are incompatible with metal containers and cause corrosion and deterioration of the metal. Others are toxic and must be handled with extreme care. The most severe problem with eutectic salts is their short-lived reversibility. Many will stop cycling from solid to liquid and back after many uses, or form solids which become permanently deposited on the bottom of the storage tank. Some eutectic salts react violently when exposed to air or moisture. Many advances are currently being made in their development to eliminate the problems which prevent them from being universally used as a heat storage material.

solar powered cooling

In most areas of the United States solar energy can also be used to cool buildings. It can be used directly to run an absorption refrigeration system or a Rankine cycle cooling system. In areas where there is a significant difference between wet bulb and dry bulb temperatures, the sun can be used in passive systems to ventilate, cool, and temper the air. This capability will be further addressed in subsequent chapters.

Heat pumps requiring a small amount of electricity can be used to heat (solar assisted) and cool buildings. Under certain conditions, the amount of electricity they require is less than the amount required to run conventional air conditioning units. For heating they require about half the amount of electricity as would resistance heating for the same output. The Coefficient of Performance (COP) of a heat pump operating in the heating mode is the ratio of the useful heat delivered to the building over the heat equivalent of the energy input to the heat pump. Even without solar assistance, some heat pump systems attain a COP of 2 or more, using earth, water, or air as low temperature heat sources.

Absorption systems which use heat from a solar collector to run their generator require much equipment and piping, and involve a large initial investment. Systems using the Rankine cycle for cooling use solar energy to power a turbine which operates a compression refrigeration unit in the system. Both of these systems can be used to remove heat from the air. The cooled air is then distributed throughout a building.

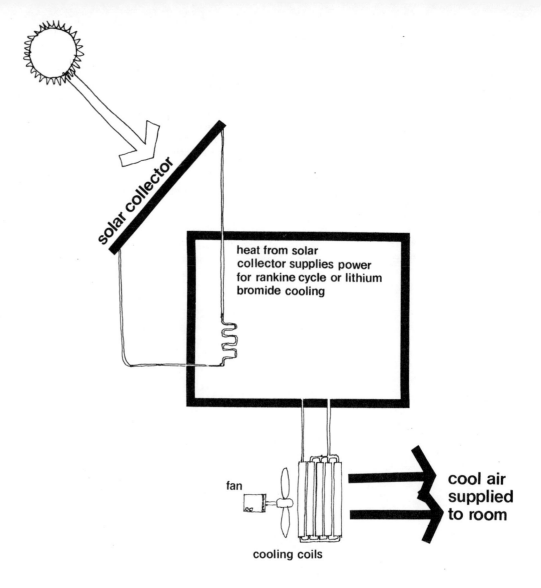

solar collector

heat from solar collector supplies power for rankine cycle or lithium bromide cooling

fan

cooling coils

cool air supplied to room

solar pond

Another device for the collection of solar radiation is a solar pond. Basically, this is a shallow body of water which itself is an energy absorber, and can be fitted with a cover plate if desired. The temperature of the water is increased by the direct absorption of photons or by conduction from the bottom of the pond. The amount of energy absorbed by the bottom is increased if it is lined with black plastic or some other black material. As the temperature of the water increases, the water expands, decreasing its density. The hottest water will be the least dense and will normally rise to the surface. Once in this location the heat is readily lost to the outside air, by conduction and convection, making this an inefficient collector. In most ponds the cover plate, fabricated from clear plastic, is used to keep it free of dirt, leaves, and other debris, and is not a barrier to heat.

Another method, relying on chemical properties, is used to reduce heat loss. This produces the class of ponds known as saline gradient solar ponds. Their successful operation depends on two properties, the first being that the amount of salts which can be dissolved in water increases rapidly with rising temperature. The hotter the water gets the greater the amount of salt that can be dissolved in it. The second property is that as the concentration of salt increases, the weight per unit volume, density, increases.

For water the increase in concentration of certain salts held in solution helps to overcome the reduction in density due to expansion. It is possible to obtain a stable situation in which the hottest fluid is also the most dense and locates at the bottom of the pond, producing gradients in both temperature and salt concen-

tration, with decreasing magnitudes of each toward the surface. Salt will tend to diffuse from the highly concentrated zone at the bottom to the zone of lower concentration at the top. Periodically, salt must be removed from the surface and reintroduced at the bottom if the gradients of both salinity and temperature are to be maintained.

The advantage of these ponds is that because the saline gradient induces the hottest water toward the bottom, convective energy losses are reduced, increasing the amount of heat that can be extracted from the pond.

For small ponds, insulation is needed to decrease the amount of heat lost to the ground. In many designs the greatest losses would occur through this region.

Heat exchangers, located at the bottom of the pond, are used to extract the heat, which can be used to heat interior spaces or domestic water, as well as power absorption cooling equipment. In experimental ponds in Israel, equilibrium temperatures have been found which approach 100 C (212 F).

Certain problems are encountered with solar ponds which decrease their efficiency, and hence their desirability. In the winter, snow and ice may cover the pond, blocking incident radiation and reducing water temperature. Even if covered, ponds need to be cleaned periodically to decrease their opacity. The cleaning process disturbs the thermal and saline gradients which take time to re-establish.

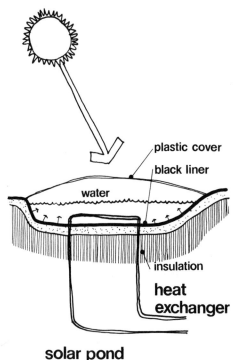

plastic cover
black liner
water
insulation
heat exchanger

solar pond section

thermosiphon collector system

Probably the simplest solar water heaters are ones that operate on the thermosiphon principle. These consist of an insulated storage tank, flat plate water type solar collector, and various pieces of insulated pipe. The bottom of the storage tank is situated at least 30 cm (1 ft) above the top of the collector, and must be high enough above it that cold water will flow by gravity from storage to the collector base. Water is heated in the collector, and because hot water has a higher specific volume than cold water, it will rise to the top of the collector, and back to the upper portion of the storage tank. As the hot water flows up the collector and back to the top of the tank, cold water is drawn from the bottom, setting up a continuous circulation of water.

This action will continue as long as there is solar radiation on the collector. The temperature of the water in the upper part of the tank will vary from approximately 74 C (165 F) on hot summer days to about 46 C (115 F) on cold winter days, unless it is too cold for the system to be operating. This type of system is well suited to preheating water before it enters a conventional hot water heater.

The storage tank, feed and return pipes, outlet pipe, and the back of the collector must be well insulated. In most warm climates one pane of glass covering the collector is sufficient, but for colder climates two sheets will be necessary. For the best flow, the cold water feed pipe to the collector should be continuously sloped from the storage tank to the base of the collector, avoiding horizontal sections.

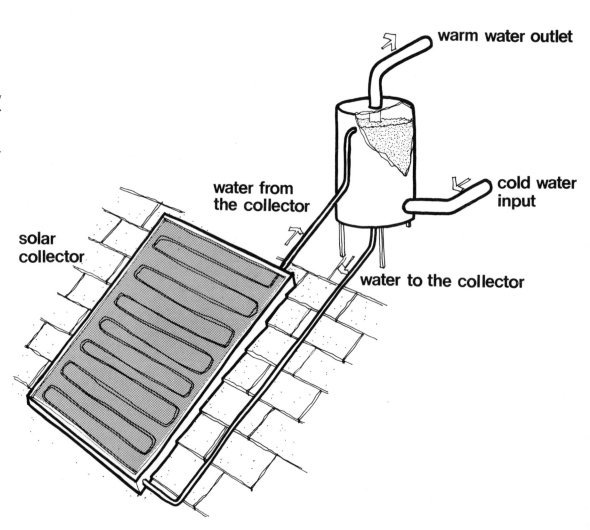

warm water outlet

cold water input

water from the collector

solar collector

water to the collector

solar domestic water heating

One of the fastest growing uses for solar energy is the heating of domestic hot water. Most collectors for home heating can accommodate the addition of a hot water preheat cycle to raise the temperature of incoming city water to a higher temperature before it goes to the hot water heater. This preheat leaves the hot water heater with a smaller difference to make up between the temperature of the incoming water and of the outgoing "hot" water, conserving "makeup" heat.

If incoming city water is at 4 C (39 F), and the hot water use temperature desired is 60 C (140 F), then the hot water heater must raise the water temperature 56 Celsius degrees (101 Fahrenheit degrees). Since each pound of water requires one Btu to raise its temperature one degree Fahrenheit, 101 Btu per pound are needed to raise the water temperature to 60 C (140 F). The addition of a solar-powered hot water preheat unit to the system could raise the temperature of the water to 38 C (100 F) before it enters the hot water heater. With the incoming water to the hot water heater at 38 C (100 F) and outgoing still 60 C (140 F), the temperature makeup is only 22 C (40 F), and will require only 40 Btu per pound of water to raise the temperature to 60 C

(140 F). In the summer, substantially higher pre-heat temperatures can be achieved, and little or no auxiliary heat is needed.

In areas with moderate temperatures all year long, such as southern California and Florida, solar heating of homes is rarely necessary, but solar heat can be economically used for heating of domestic water. The largest number of solar hot water heater manufacturers and users are found in these geographic areas. The size of a solar collector used for water heating only is much smaller than one used for both space heating and domestic water heating. Solar water heaters can also be used to heat swimming pools and whirlpool baths. The collection system for hot water heating is similar to the flat plate liquid collector system, and most operate on the thermosiphon principle.

The size of any storage tank and collector system depends on the initial cost, the rate of hot water use, the temperature to be maintained, and period of time during which there occurs minimum sunshine. Some collectors incorporate their own storage systems, recirculating and reheating the water in a closed loop. The collector can range from simple black plastic bags on the roof to moderately complicated collectors with automatic temperature controls and heat exchangers to maximize the energy use. In climates where freezing occurs, self-draining collectors with insulated tanks are necessary to avoid damage to the system.

solar collector

heat exchanger

hot water preheat tank

hot water heater

hot water

incoming cold water

solar water distillation

Solar powered distillation (purification and deminer-
alization of water) and desalination (removal of salt
from water) projects are possible using the sun's
energy to evaporate and purify water supplies. One
simple system has a glass cover over the water
supply. The cover is sloped toward a collection
point or pipe. Usually the surface below the water
is colored black to absorb the maximum possible
incoming radiation. As the space enclosed by the
glass heats up, the water temperature increases and
its surface evaporates, leaving behind the minerals
and salts that were dissolved. The water will condense on the lower side of the glass cover, which,
due to its exposure to outside air, is cooler than the
heated chamber. As the water droplets form, they
will merge together and then flow to the lower end of
the cover and be collected as distilled water, which
may then be bottled or otherwise distributed. The
minerals and salts which are deposited on the bottom
of the still must be cleaned away periodically.

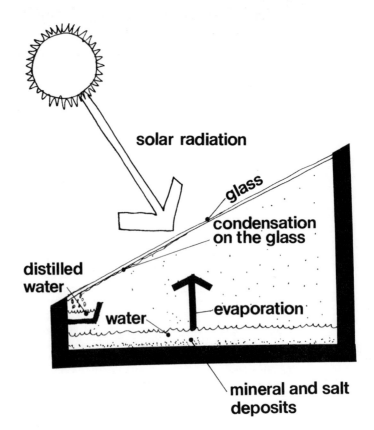

solar radiation

glass

condensation
on the glass

distilled
water

water

evaporation

mineral and salt
deposits

large scale solar collection

Solar collection can take place on a large scale to generate electricity or provide steam to heat a district. To generate electricity high temperatures are required. These can be derived using collectors with reflectors and concentrators.

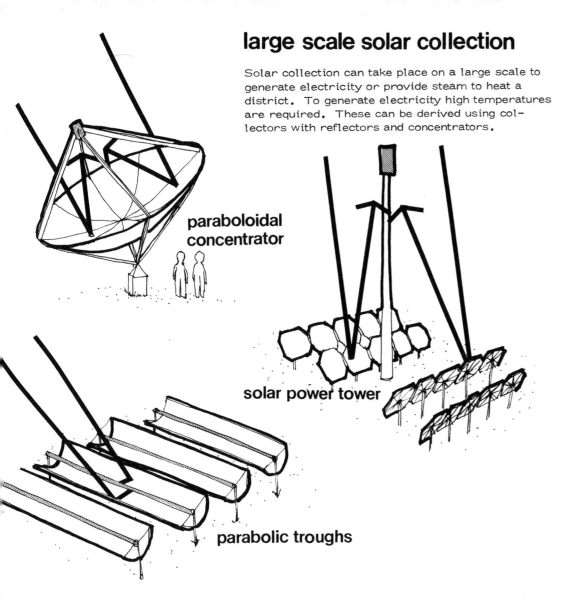

paraboloidal concentrator

solar power tower

parabolic troughs

Listed below are three proposed schemes for using solar energy on a large scale:

1. Use an array of parabolic reflectors to concentrate solar energy on small focal areas to produce high temperatures for steam production and turbine operation.

2. Focus the sun's rays on piping with an array of linear/parabolic trough reflectors. Water run through the piping will be turned to steam by the high temperatures. The steam under pressure can be transported over long distances and be used in large power generation plants.

3. The "power tower" arrangement uses a focal point tower 75 to 100 meters high situated among several thousand flat mirrors which track and reflect the sun to the focal point of the tower. Steam is then generated in the tower.

These systems would convey steam to generators which would then provide electricity.

Another proposed, yet currently cost prohibitive, system for large scale generation of electricity from solar radiation would use photovoltaic cells. Large scale operation would work similarly to smaller home systems but batteries would prove impractical in terms of cost and materials. The energy collected could be best utilized as mechani-

cal potential energy, such as pumping water up to
a higher level when energy use is low and then
allowing it to flow to generate electricity for high
demand periods. Also possible, yet expensive, would
be energy storage by the electrolysis of water, producing hydrogen and oxygen, and their subsequent
recombination in a fuel cell.

To make solar heat collection systems cost effective
in a building, energy conservation must be made a
top priority. New designs and forms will result in
a climate-responsive ethic which may soon override
our more familiar historic architectural motivations.

Typical urban and suburban houses are major energy
wasters and most cannot economically facilitate the
integration of solar energy. Some manufacturers
market solar systems on the basis that one can add
these units to the exterior of a home without any
additional changes to the building. This is uneconomic
unless energy conservation is made a primary effort.
Without the use of energy conservation measures,
any form of energy, whether solar or fossil fuel, is
poorly and wastefully utilized.

Wind/air

Wind and air movement are the results of the sun's warming of different layers of the atmosphere and parts of the earth's surface. As the sun strikes the earth, areas near the equator are heated more than areas near the poles. This causes equatorial air to rise to upper levels of the troposphere (the atmospheric zone closest to the earth's surface). Cold polar air moves to the equator to replace the air that has risen, and sets up circulating planetary wind patterns. The rotation of the earth has an effect on these winds, in what are known as Coriolis forces (after the French civil engineer who identified them in 1843): cooler air moving along the earth's surface toward the equator is diverted toward the west, while the warmer air at the upper levels of the troposphere tends toward the east. These effects initiate large counterclockwise circulation of the air around low pressure areas in the northern hemisphere, and clockwise circulation in the southern hemisphere. Topological features, day-night heating and cooling cycles, and man-made structures further complicate airflow patterns and generate microclimatic winds.

Localized wind movement and direction depend on the location of high and low pressure areas (equalization of pressure areas), the density of air (barometer readings), the amount of cloud cover (temperature differentials caused by shade or sun), topography (shape and contour of the land), land/water formations, and the position and tilt of the earth with respect to the sun. All of these factors affect wind in such a manner as to generate a much more complicated pattern than is described by the simplified cell diagram. It shows, however, that basic air movement patterns are caused by the sun.

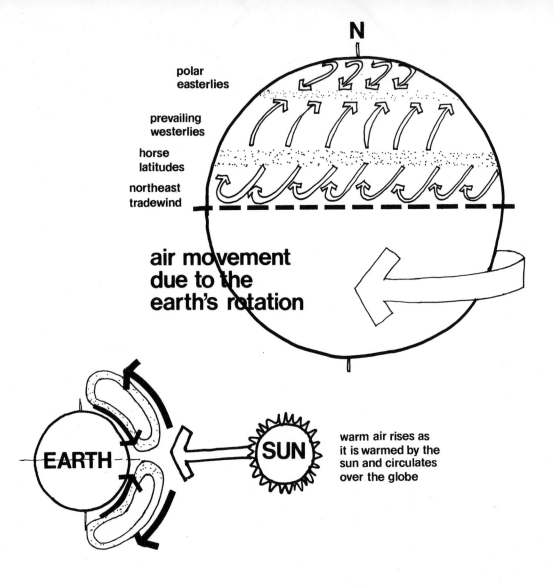

N

polar
easterlies

prevailing
westerlies

horse
latitudes

northeast
tradewind

**air movement
due to the
earth's rotation**

EARTH

SUN

warm air rises as
it is warmed by the
sun and circulates
over the globe

Man has utilized wind energy for many thousands of years but has never taken advantage of its full potential. Wind has not been used to propel large ships since the early 1900's, although its use for the propulsion of sailing vessels dates back to the Egyptians, circa 3000 B.C. The development of the steam engine and other mechanical systems gradually supplanted the use of wind energy to propel large sailing vessels.

Wind can propel many types of vehicles; sails are even available for bicycles. The movement of the air exerts a pressure on any surface with which it comes in contact. With a constant wind speed the force acting perpendicular to a surface increases at the same rate as the area of the surface increases. The surface can either be stationary such as the side of a building, which must transfer the force of the wind to the ground; or movable, and be set in motion by the force of the wind. Wind will exert a pressure of approximately 2394 newtons/square meter (50 lbs/square foot) on a surface, at a wind speed of 240 km/h (150 mph).

Wind energy may be harnessed for the movement of a sail or a blade. Many old windmills (for milling grain) in Europe used sails stretched over wooden frames to generate mechanical movement. The movement of the rotating sails was then transformed to useful work by wooden gears and shafts. Windmills are generally thought of as producing direct mechanical energy while wind generators and wind turbines produce electrical energy. The majority of the wind machines being developed today are for the generation of electricity, yet most units presently operating pump water.

71

The theoretical maximum amount of kinetic energy which can be extracted by a wind machine from the windstream is only 59.26 percent. This factor is known as the Betz Coefficient, first derived in 1927 by the German engineer A. Betz. His calculations assumed 100 percent efficiency within the wind machine system itself. Since no machine can be perfect in construction and operation, and since energy is lost in electrical transmission lines, only about 40 percent of the wind's kinetic energy can be practically utilized in buildings.

The amount of available wind energy varies depending on the location, elevation, and orientation of a wind machine. Climatic data have been collected on the force of wind and its direction for many areas of the United States. For some locations the wind speed and direction are plotted on a polar axis graph (graph with North, South, East, and West indicated) in terms of wind strength and frequency. The resulting plot is called a wind rose map, in reference to the shape most often generated; percentages indicate the portion of time (annually in this example) during which wind speeds reach the figures shown. Also helpful in determining the approximate amount of power which can be extracted from the wind are calculations for monthly average wind speeds and annual average velocity duration curves.

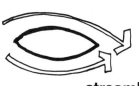

streamlined

irregular

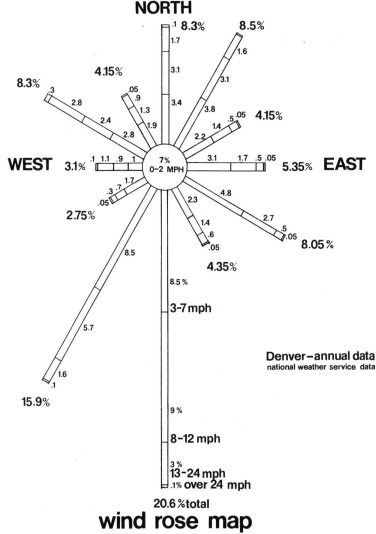

Denver–annual data
national weather service data

wind rose map

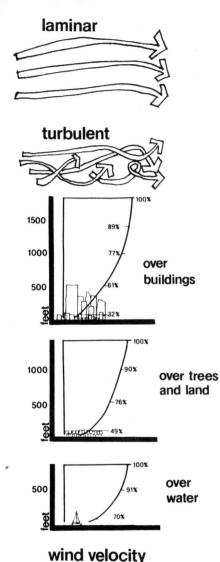

laminar

turbulent

over buildings

over trees and land

over water

wind velocity profiles

The quality of wind also affects the amount of power that can be extracted from it, as illustrated. There are significant amounts of shear and compression in a horizontal windstream flowing over the surface of the earth. This shear results in lower wind speeds close to the earth's surface than at altitudes at which free flow occurs. Laminar flow is the movement of air in parallel layers over an even landscape or calm body of water. An interruption in topography causes turbulence in the wind pattern. A tree, a house, or a hill can cause multiple redirections of wind in a manner which would cause instabilities in a poorly located wind machine.

Therefore, it is usually suggested that wind machines be placed as high as possible off the ground to eliminate surface effects. A customary minimum distance is ten meters (33 feet) from ground level to the bottom of a turbine or airfoil. In nearly all geographic locations, winds become laminar above this height, and their velocity increases sharply. (The power of the wind increases as the cube of the velocity.) For large scale wind generators, heights may reach 65 meters (213 feet) or more.

Great care must be taken in choosing a suitable site for a wind machine. It is possible to increase the average power output by siting a machine to take advantage of local topographical features. Flow is accelerated as it passes over a rounded hill or through a narrow valley. Because of local anomalies, it is worthwhile to make a detailed wind survey before choosing a site for a wind machine.

Wind machines convert the horizontal force of wind into rotary or oscillatory mechanical motion. Limitations on the mechanical efficiency as mentioned above restrict the amount of energy which can be removed from the wind. One important principle in the design of wind machines is the tip speed ratio, the relationship between blade tip speed and actual wind velocity perpendicular to the plane of rotation. The higher the tip speed ratio is, the more efficient the machine will be. Many old style windmills produce tip speed ratios of 1 to 2 and develop more torque (turning power) than is necessary for electrical generation. The most recently developed wind turbines and generators usually feature tip speed ratios between 6:1 and 8:1 facilitating more efficient electrical generation than is possible with lower tip speed ratios.

Wind machines produce useful power between certain minimum and maximum rotational speeds. Most machines need at least 12-16 km/h (7.5 – 10 mph) wind to generate electricity. Older windmills can use wind speeds as slow as 3 to 4 km/h (1.87 – 2.5 mph) to pump water. The top utilization speed of wind machines averages approximately 65 km/h (41 mph). No additional useful power can be extracted, without adjustments, from winds above that speed. Some wind machines will let the turbine, blades, or rotors spin on a free clutch system or centrifugal governor in high velocity winds, while others have blades which can be mechanically modified to maintain a constant rotational speed in variable winds.

Much research has recently been done on airfoils, sailwings, hoops, rotors, single propellors, paired contra-rotating propellors, and turbines to determine their suitability as part of a wind machine system. Each of these mechanisms may be categorized as either a horizontal or vertical axis device.

Horizontal axis machines are typified by old windmills, in which the plane of rotation is, ideally, always perpendicular to the prevailing wind direction. These machines must be equipped with a mechanism to turn them when the wind direction changes. This can only be accomplished via the tracking process in which axis and blades all pivot into the wind. Because of tracking, some mechanical energy is lost. Horizontal axis devices have nonetheless enjoyed popularity for centuries, as they require only a low wind speed for fairly successful operation.

Horizontal axis wind machines vary greatly in their performance and design. Wind turbines consist of many blades which convert wind power into rotary motion. They can have an electrical generator connected to the central shaft or one positioned to ride the outside circumference of the wheel, as shown in the illustration. The speed of the central shaft is generally too slow to power a generator, without gearing upward at some cost in efficiency. By contrast, generators operating on the outer circumference of the wheel can take advantage of the high velocity of the blade tips. Shown is a variety of horizontal axis wind machines.

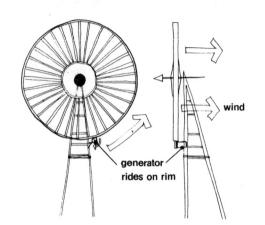

bicycle wheel turbines

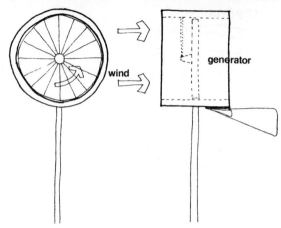

gillette wind turbine

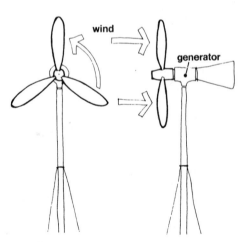

3-propellor wind machine

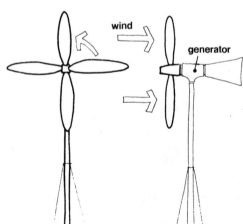

4-propellor wind machine

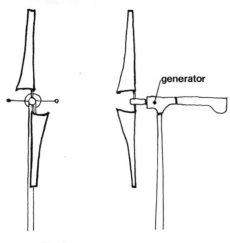

sailwing rotor

The rotor of a vertical axis wind machine revolves in a horizontal plane. Because of this orientation the machine does not have to pivot to face into the wind. This feature is a distinct advantage on wind machines being used in areas where sudden changes in the wind direction occur, as high stability is exhibited and little "adjustment energy" is lost to any wind direction or velocity change. Therefore, wind gusts would cause power surges rather than disruptions.

Vertical axis machines have been used intermittently for centuries but have been less applicable to small installations than to large scale milling operations and water pumping.

Many vertical axis wind machines produce high torque at a low speed. This is useful for operating an irrigation pump but is too slow to directly power an electrical generator. Some wind machines, of both the vertical and horizontal axis variety, use gearing mechanisms to convert high torque, slow speed rotary motion into low torque, high speed motion sufficient to power an electrical generator. Advancements have recently been made in the development of electrical generators which operate at low speeds.

The two most established types of vertical axis wind machines are the Darrieus hoop rotor and the Savonius rotor (also known as the S-rotor).

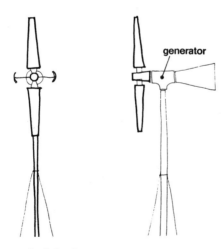

2-blade rotor

The Savonius rotor was patented in 1929 by Finnish engineer S. J. Savonius. This is predominantly a drag device though it does provide some lift. The vanes of the rotor catch the wind, striking the back of the vane and exiting behind the rotor shaft. A single rotor can have more than two vanes, in which case the torque may become more constant and the axial balance more refined. Another configuration accomplishing this result, shown at far right, is the stacking of several two vane S-rotors at different angles, along a central shaft. The S-rotor operates at a maximum efficiency of 31% with a tip speed ratio of 0.8 to 1.8. The S-rotor can be economically built from oil drums cut in half vertically, then welded between end caps to form scoops; these in turn are mounted on a central shaft and the shaft connected to a diaphragm pump or electrical generator.

The Darrieus hoop rotor was patented in 1931 by the Frenchman G.J.M. Darrieus. Research has been conducted recently on the rotor by NASA in the USA and by the National Research Council of Canada. Both teams have found high potential efficiency in this configuration, within a certain windspeed range. The blades of the rotor are flexible and bend in response to the wind. Darrieus rotors are lift devices, which are characterized by curved blades with airfoil cross-sections. They have relatively low starting torques, but relatively high tip-to-wind speeds and high power output for a given rotor size, weight, and cost. Usually the rotor needs some type of starter system to initiate its rotation. Frequently, a small S-rotor is mounted at the base of the central

shaft to start the Darrieus rotor spinning. Such an addition increases the weight and cost of a system, so trade-offs between maximum power output, starting efficiency, and cost must be considered in developing an optimum design for a given application. The efficiency of the Darrieus rotor is approximately 35% with a tip speed ratio of 6 to 8, depending on the type of rotor. The main advantage of this type of system is its low cost.

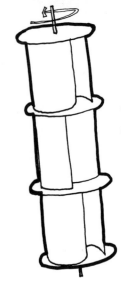

three S-rotors stacked

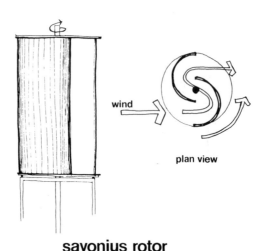

wind

plan view

savonius rotor

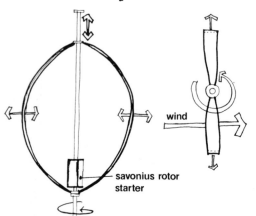

wind

savonius rotor starter

darrieus hoop rotor

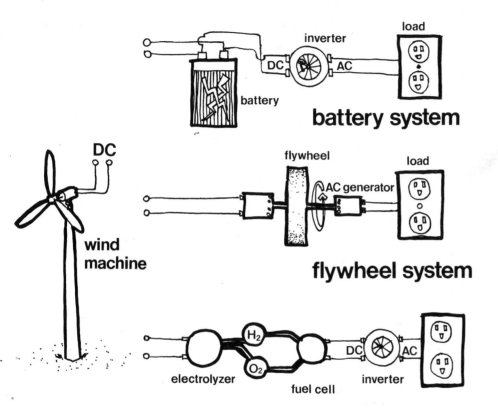

battery system

flywheel system

DC

wind
machine

fuel cell system

energy storage

One of the main problems facing the use of wind for small scale electrical generation is the present limited ability to store the energy. Batteries are the most efficient chemical storage system now available, but banks of batteries are costly to install and to maintain. The power generated by a wind machine is not constant enough to guarantee a reliable quantity of electricity or a stable AC frequency. Therefore, most wind generators produce DC (direct current) voltage. Electricity in this form is compatible with storage in batteries, which themselves yield only DC voltage. Resistance appliances such as heaters, toasters, and incandescent light bulbs can be operated on DC current, but most large appliances contain induction motors which will operate only on AC. It is possible to use an inverter to change DC to AC, although these devices are quite expensive. Some authorities foresee a possible resumption of the general use of DC motors to power large appliances.

Two of the most promising developments for wind energy storage appear to be the flywheel storage system and the electrolysis/hydrogen fuel storage system.

A flywheel can store a much higher quantity of energy per mass weight than a battery. The principle of the flywheel system is that a spinning wheel can store and accumulate energy as momentum and release it for later use. Advances in material technology and low friction bearings have made the flywheel principle feasible for energy storage. Flywheels can be mounted in a vacuum to decrease air friction. As the mass of a flywheel increases and the speed at the rim increases, there is an increase in the amount of stored energy.

The amount of energy that can be stored in a flywheel is a function of the material from which the
flywheel is made, its size and shape, and its speed
of rotation. Some heavy materials develop greater
internal stresses than lighter materials at a given
speed of rotation. But lightweight flywheels must
be spun faster to store the same amount of energy
as the heavy ones. A multi-ring or radial fiber
composite material flywheel could store 30-40
times as much energy per pound as a lead-acid
battery. The flywheel would be attached to a motor/
generator unit to recover the energy.

A motor powered by electricity from a wind machine
can spin a flywheel and continue to add energy as
long as the wind blows. When the wind speed is
sufficient to generate continuous electricity, the
motor can turn off and the flywheel will continue to
spin. When electrical needs could no longer be met
directly by the wind generator, the flywheel could
then spin its own generator. It is estimated that a
flywheel sealed in a partial vacuum and riding on
extremely low friction bearings (possibly magnetic
suspension) could store energy for months, though
this has not yet been demonstrated.

The electrolysis of water to produce hydrogen and
oxygen is another alternative for energy storage,
in which electricity generated by the wind machine
breaks down the water into its two components,
hydrogen and oxygen. These two gases are then
stored in tanks for future use. Energy is released
as electricity in a hydrogen fuel cell, where the
hydrogen and oxygen are recombined in giving off
water vapor as they produce electricity.

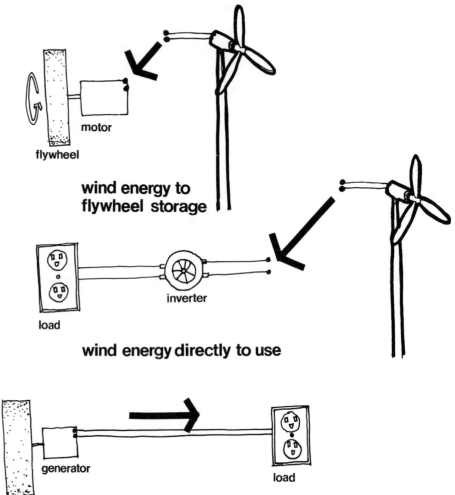

flywheel

motor

**wind energy to
flywheel storage**

load

inverter

wind energy directly to use

flywheel

generator

load

flywheel stored energy to use

large scale power generation

The use of wind energy is by no means limited to small scale applications. It is theoretically possible to build wind generators that can deliver several megawatts of power. There are certain constraints on the size and power output of wind machines because of limitations on the strength of materials, which in turn limit the allowable stresses for blades, bearings, towers, and other components of the system. Cost is also a limiting factor. In certain applications it may be cheaper to build two small machines rather than one large one of equivalent total output.

The most effective application of wind energy would be to use it directly to pump water for irrigation, or to generate electrical power. Wind systems without energy storage could be used in conjunction with a fossil fuel powered generating plant to save fuel when the wind is blowing. However, for most applications energy storage would be required so that uninterrupted power could be delivered. Storage could be accomplished by the use of pumped water systems, compressed fluids, stored electrolytic hydrogen, flywheels, and other forms mentioned above.

Three types of wind machines would be suitable for large scale applications: those having open horizontal axis rotors, open vertical axis rotors, or vortex generators. (Small scale vortex generators may also be practical in the future.)

Wind machines with horizontal axis rotors for large scale applications are similar in form to the ones already discussed for small scale applications. In a 27 kilometer per hour (17 mph) wind, a machine with an 18 meter (59 foot) diameter rotor has a rated output of 100 kilowatts (kw). One with a 50 meter (164 foot) diameter rotor is rated at 1.0 megawatt (mw), while one with a 136 meter (446 foot) diameter rotor is rated at 10 mw, but would entail the height of a 45 story building in addition to its mounting height. The safety of life and property in the vicinity of a wind machine of such scale is a primary design consideration. In case of a rotor failure, buffer zones would have to be established in which occupancy is limited. Ultra-large machines could be placed in remote areas, or in the ocean, where structural rigidity would not be as critical as on land. Smaller machines could be used near populated areas, and possibly arranged in sizable arrays.

Vertical axis wind machines of the Darrieus type could also be used for large scale applications. In a 27 kilometer per hour (17 mph) wind, a rotor with a 21 meter (69 foot) diameter rotor has a rated output of 100 kw. One with a 58 meter (190 foot) diameter rotor is rated at 1.0 mw, while one with a 158 meter (518 foot) diameter rotor is rated at 10 mw. It is also necessary to take safety precautions and properly locate this type of machine, although sudden wind reorientation is far less a challenge than with horizontal axis turbines.

Another concept in wind machines is the vortex generator. These machines spin the wind to increase the power output of a turbine located in or near the vortex. Two varieties are presently under study. They are the unconfined and confined vortex type. One type employs wing-like structures to deflect the wind and create an unconfined vortex around the turbine. It is estimated that an unconfined vortex generator can be designed to provide up to six times the power output of a conventional system, with the same diameter rotor.

Grumman Aerospace Corporation is developing a confined vortex system in which the pressure drop across a ducted turbine and the wind velocity through it are augmented by using additional ambient wind to produce a confined tornado-like vortex in a tower located at the exit of the duct. For a typical confined vortex system, the diameter of the tower might be three times the diameter of the turbine, and the height of the tower might be three times its diameter, or nine times the diameter of the turbine. Dr. James Yen of the Grumman Aerospace Corporation estimates that large scale wind energy systems, which have ducted turbines interconnected to vortex generators, may be designed to have power outputs that are 100 to 1000 times those of conventional systems, of the same rotor diameter, operating at the same ambient free-flow wind speed.

With these systems, smaller high speed turbines can be used to obtain the same output power as the large bladed conventional systems, thus avoiding the large weight and inherent stresses on the blades of the conventional systems.

These concepts can yield significant improvements in the design of wind machines. They indicate that high power output can be derived from relatively small installations, and that the structure needed to deflect the wind can also shroud the turbine, decreasing potential dangers.

The cost of generating electricity with wind-vortex systems can be markedly reduced by incorporating the vortex tower in a multi-purpose building.

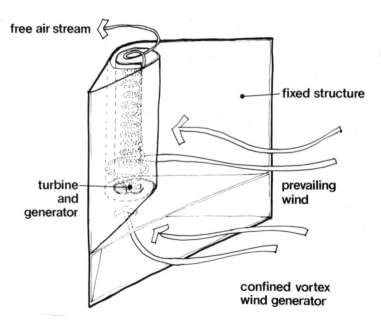

free air stream

fixed structure

turbine and generator

prevailing wind

confined vortex wind generator

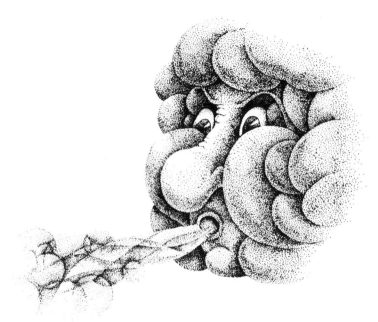

There are a number of additional ways to store the energy produced by large scale wind machines; most require interconnecting the machines with other systems.

In pumped water storage systems, the wind machines are used in conjunction with a hydro-electric system. Water from below the dam is pumped into storage behind it, using wind power. This can be accomplished with direct mechanical pumping or with electric powered pumps. The water behind the dam flows through turbines, generating electric power, and is retained in a holding basin. It is then pumped behind the dam again, using wind machines, and the cycle starts over. With this type of system a steady supply of electric power can be supplied.

While wind machines are among the simplest energy conversion devices, the wind is an unpredictable force. The selection of an optimum site for a wind machine and the choice of optimum designs for particular applications are often difficult and complex problems. Much research and development is currently underway on wind machines; undoubtedly, innovative designs will continue to emerge. Wind machines, varying with geographic location, appear to have a high potential for making significant contributions to the goal of meeting our future energy needs.

Wind energy is derived from the sun's energy, and its use has no discernible ill effects on the environment. Harnessing the wind's energy produces no thermal pollution, no air pollution, no radioactivity, and no other by-products which are dangerous to populations.

notes

Water/
precipitation

There are three major forms of water energy on earth: gravitational, tidal, and thermal gradient. Ocean water moves due to constant change within the hydrologic cycle, and the effects of the earth's rotation, as well as lunar influence (in larger bodies of water). Evaporation from bodies of water brings about cloud formations which redistribute the water to higher land elevations in the form of rain and other types of precipitation. This water then flows into streams, creeks, then rivers, and back into the ocean. The sun provides the energy for evaporation and condensation in clouds, also propelling the clouds and their inherent payload of water to new locations.

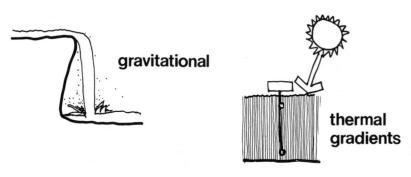

gravitational

thermal gradients

tidal power

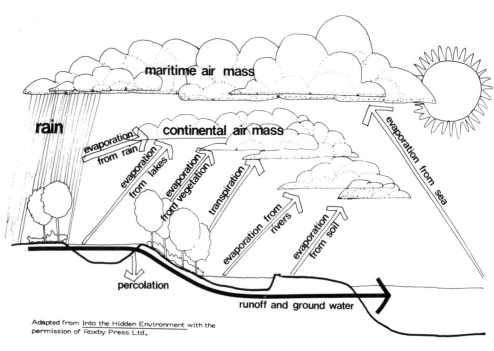

maritime air mass

rain

continental air mass

evaporation from rain

evaporation from lakes

evaporation from vegetation

transpiration

evaporation from rivers

evaporation from soil

evaporation from sea

percolation

runoff and ground water

Adapted from Into the Hidden Environment with the permission of Roxby Press Ltd.

hydrologic cycle

Electrical power generation using the gravitational potential of water is the most developed method of extracting energy from the hydrologic cycle. The earliest known use of water power was to drive pumps for irrigation and power mills for grain processing. Later, water wheels were used to power textile plants in Europe and the United States. The wheel would supply power to a central shaft and by means of belts and pulleys, the energy was transferred to individual machines.

The two most common types of water wheels are the undershot wheel and overshot wheel. The undershot wheel can be installed adjacent to or over an existing stream. Paddles or buckets on the outer rim of the wheel are propelled by flowing water, developing a turning or torque power of the wheel.

The overshot wheel needs a cliff, dam, or sluice to raise the level of the water behind it to the top rim of the wheel. Generally a sluice, a small channel, is set up wherever terrain prohibits the more natural forms of channel. As the water leaves the sluice, it drops into the scoops on the wheel. The weight of the water forces the scoop down, turning the wheel.

Both the overshot and undershot wheels work at slow speeds delivering high torque power useful for mechanical purposes.

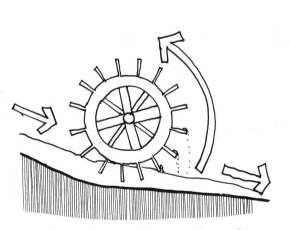

undershot wheel

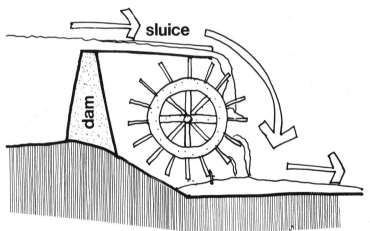

overshot water wheel

Another system using water motion is the turbine, which can develop much higher rotational speeds than water wheels and is used for generation of electricity. With the turbine, water is forced through a pipe or tube. As it exits the tube under pressure, it strikes the blades of the turbine causing it to turn. Most hydroelectric dams use large scale turbines to generate electricity. To develop the water pressure necessary to operate a turbine, a dam is needed to develop a vertical column of water or pressure "head".

When a river or stream is dammed, the ecological impacts of the dam must be taken into consideration. The dam interrupts the flow of water and prevents fish and other water creatures from reaching areas downstream and upstream. The water behind the dam will evaporate at a much higher rate because of its changed surface area to volume ratio. These extra pressures can result in shifts in geological formations. New vegetation may start in areas adjacent to the banks where high water level conditions did not exist before. In areas where the amount of water varies due to precipitation changes, reservoirs can become mud ponds during periods of little precipitation and the accompanying low water levels.

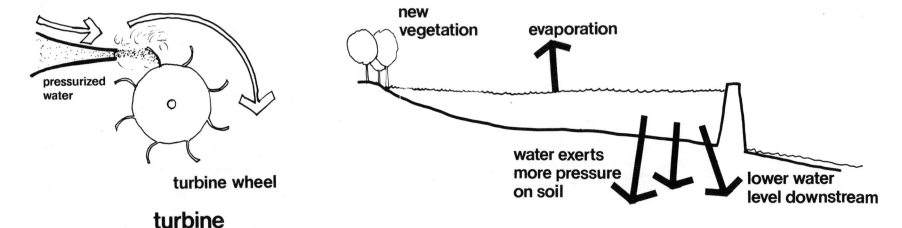

pressurized water

turbine wheel

turbine

new vegetation

evaporation

water exerts more pressure on soil

lower water level downstream

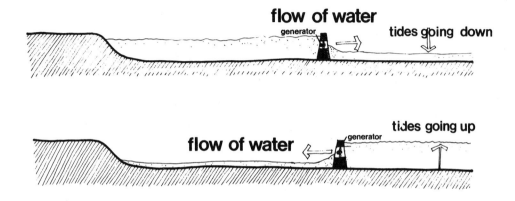

flow of water

generator

tides going down

flow of water

generator

tides going up

tidal power

The earth's tides are not directly related to solar energy but are another force of nature. As the moon travels around the earth, gravitational force between the two acts on the oceans. The effect of the moon orbiting the earth and the earth revolving on its axis produces high and low tides on the earth's surface. The motion of this water represents a tremendous amount of power.

Generators mounted between or within dams could be constructed to economically convert this power into useful energy - electrical power, for example. Such a system would trap alternately high and low tides as illustrated; height of water head and actual horizontal motion would combine to offer extremely large-scale cyclic power potential.

Thermal gradients are differences in temperature at varying depths; in natural bodies of water other than geothermal ponds, these gradients are a result of heating of upper water levels by the sun. As the sun radiates to the water, currents are induced (similar to wind currents) in which warm water rises to the surface then flows to cooler water areas where it drops in temperature and, flowing to the bottom, completes this cycle. As with wind, this is an over-simplified rendition of changes which, interrupted by turbulence, actually occur. Currents are affected, for example, by variations in the ocean floor terrain, geological irregularities producing geothermal warming, and by air currents above the ocean.

Low levels of the ocean are generally colder than areas near the surface. These differences in temperature can provide power for Rankine cycle generators as illustrated.

Since the earth itself is a solar collector, and the majority of the earth's surface is water, the oceans are a most effective means of collecting and storing the thermal energy of the sun. The temperatures in the ocean vary as to the amount of localized cloud cover and the measurement of depth below the surface. Tropical ocean temperatures may range between 4 and 27 C (39 and 81 F), suggesting, in some areas, the existence of strong convective currents.

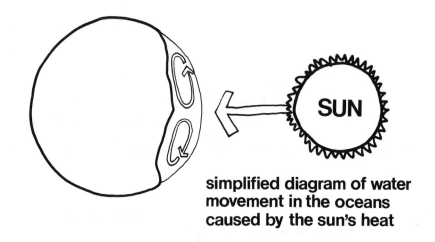

simplified diagram of water movement in the oceans caused by the sun's heat

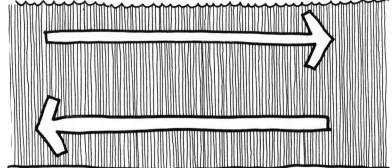

warm water currents

cold water currents

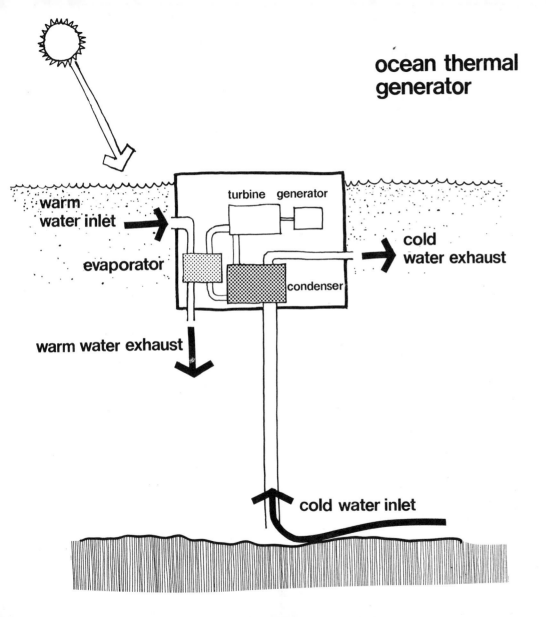

ocean thermal generator

warm water inlet

evaporator

turbine   generator

condenser

cold water exhaust

warm water exhaust

cold water inlet

The best collection areas lie in temperate waters between the Tropic of Cancer and the Tropic of Capricorn. In these regions also exists the highest ratio of water area to land area: nine to one. The water surface temperature here averages approximately 27 C (81 F).

Large scale electrical generating plants could utilize the temperature differentials with power generators located just below the ocean's surface. The condenser and boiler units could be several hundred feet below the surface while the generator and turbine could be at a minimal depth (deep enough for clearance below ships and still accessible to divers). Stationing of the generator below the surface minimizes danger from storms and their climatic disturbances. Water intake points should be located where maximum temperature differentials occur. The cold water entering the condenser might be found as much as 610 meters (2000 feet) below the surface, with the warm water inlet being located considerably above it. Naturally occurring differences in temperature and pressure over such a distance would cause refrigerant fluid to power a turbine. Several refrigerants are presently being considered: propane, Freon, and ammonia. Approximately 10% of the energy which the plant would produce is needed for the pumps which maintain water pressure levels. Estimates of the construction costs of such systems range from half to the same cost of a fossil fuel plant of identical power output. A further advantage over conventional fuel plants is the abundance and renewability of ocean thermal gradient structures throughout temperate regions.

Some untried ideas for large-scale electrical power
generation involve harnessing the energy contained in
deep water waves and shoreline breakers. Waves are
initiated by the sun (temperature differential), wind
(direct force), and the gravitational attraction of the
moon (tides). They contain vast amounts of power
and have the potential of being used to generate
electricity.

An object on the surface of the water bobs up and
down with the passage of each wave, yet always
seems to remain in the same spot. The wave is not
a moving mass of water, but is a force which lifts
the water's surface as it passes along. A large
float, closed cylinder filled with air, etc., on the
water's surface will rise and fall with the periodic
wave motion. Work can be extracted from this
motion and used to operate a pump which, through
an appropriate mechanical arrangement, can be
used to generate electricity.

The power from waves breaking on the shore can
also be used to generate electricity. The upward
breaking power is first converted to mechanical
power and then to electrical power. One scheme
would be to use the breaker power to compress air,
which in turn would be used to power a turbine
generator.

Earth

The earth is constantly generating heat which is transferred to its surface. This heat comes from the earth's molten center, where temperatures in excess of 1000 C (1832 F) are attained as a result of natural decay of radioactive core materials and frictional forces resulting from solar and lunar tides, plus the relative motion of continental plates. The earth radiates 54 calories per square meter per hour (0.0199 Btu/square foot/hour). As the distance from the earth's core decreases, the internal temperature level increases. For each kilometer below the earth's surface, the temperature will rise 30 C (54 F). The average temperature at the earth's surface is 10 C (50 F), but drilling to a depth of approximately 3 km (1.8 miles) will yield a 90 C (194 F) temperature which is economically useful for space heating and many industrial uses. A depth of approximately 7 km (4.3 miles) will yield a temperature of 180 C (356 F) which can be useful in the generation of electricity with the assistance of a low boiling point fluid to power a generator. The depth at which ideal steam-electric generation temperatures (240 C or 464 F) are normally found is 9 km (5.6 miles); this is near the present limits of well-drilling technology and suggests use for only large scale power generation plants.

Irregularities in geological formations can bring about and indicate changes in the depth and location of useful geothermal temperature gradients. In some areas geothermal sources of heat penetrate the surface as geysers and hot water springs. The Geysers, an area in northern California, is known for its escaping steam. The Pacific Gas and Electric Company buys the steam and presently uses it to generate approximately 200 megawatts of continuous electricity. Other geothermal sources worldwide are now being used for power generation, industrial power, and space heating.

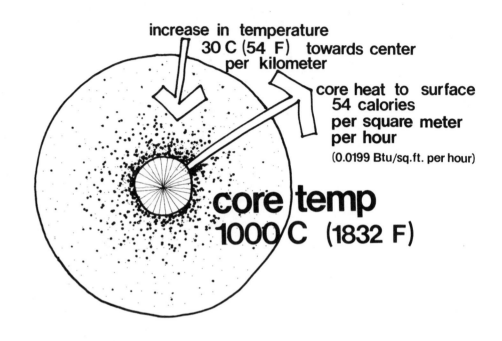

increase in temperature
30 C (54 F) towards center
per kilometer

core heat to surface
54 calories
per square meter
per hour
(0.0199 Btu/sq.ft. per hour)

core temp
1000 C (1832 F)

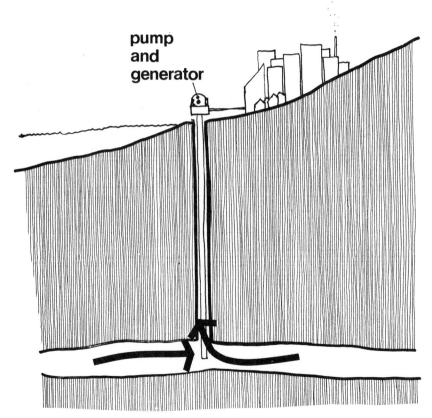

**pump
and
generator**

dry steam

geothermal

There are three major types of geothermal power:
dry steam, hot water/wet steam, and hot dry
rocks. Each has different characteristics of
depth, access, and economic feasibility. Many
of these characteristics change with each potential
site.

Dry steam is the cleanest and the most usable of the
three, but is very limited and occurs only within
unusual geological formations. Temperature levels
obtainable with dry steam are much higher than with
other geothermal sources. Dry steam temperatures
at accessible levels are approximately ten times as
high as the average corresponding depth elsewhere
should yield. Shafts are drilled into areas containing dry steam; it is filtered, then channeled for
electrical power generation.

Wet steam/hot water is more abundant than dry steam. Hydronic heating systems can use the geothermal hot water directly or remove heat through a surface-mounted or "downhole" closed-loop heat exchanger. The temperatures are sometimes high enough for electrical generation, but are usually more suited to space heating requirements. Wet steam/hot water geothermal deposits are approximately 3 times hotter than similar drilling depths.

Much of the wet steam/hot water which is available under the geographic USA contains high percentages of mineral and salt deposits. These materials can cause a number of problems with typical piping in collection and distribution systems. Hence, the closed circuit system is preferable where water composition would corrode piping. Also, there exists a number of potential hazards regarding air pollution and stream contamination when such water is exposed to the environment. If water (following heat removal) is pumped back into the underground geothermal reservoir, large craters can result when unsupported earth overlying the reservoir collapses. Several such incidents have occurred in New Zealand where water geothermal resources are widely employed.

Both hot water and dry steam geothermal reservoirs are limited in reserve quantity; estimates range from 50 to 300 years of useful life for either of these geothermal sources.

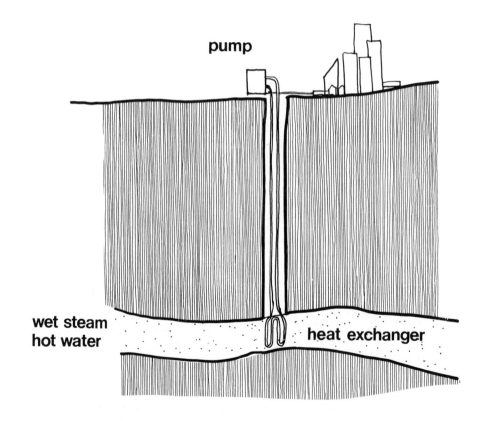

water is pumped down one hole through the rocks and then back through another hole to complete the circuit

pump

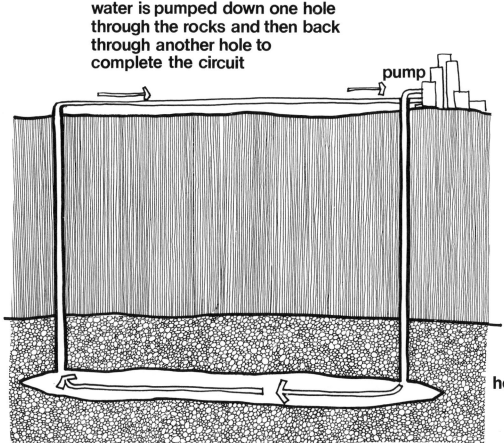

hot dry rocks

Dry rock contains the primary form of the natural heat energy of the earth, and as an energy source is practically unlimited. This heat can be removed by a number of methods. One of the simplest of these consists of drilling to a depth necessary for the temperature required, then pumping in neutralized water and returning heated water to the surface for heating domestic water or space heating.

Unless large areas in hot rock strata are fractured, the minimal exposed surface area will be insufficient for economical heat exchange. A method of correcting this, hydro-fracturing, calls for pumping the water at very high pressure into the well to crack and fracture large areas of hot rock. An auxiliary hole is then drilled to another point along the fracture completing a pipeless "circuit". Water is pumped down the first hole, is warmed by the hot dry rocks, and is piped out through the second hole. The technology for such drilling has already been developed by various oil companies.

As obstacles in engineering and economics are overcome, geothermal heat could prove to be a major energy source for the future. Techniques are being developed for large scale projects to supply space heating for housing and industry.

Another source of energy provided by the earth is gravity, that is, the attraction between any two masses. The force of attraction between the two is proportional to the product of their masses and is inversely proportional to the square of the distance between their centers of mass. As the mass of either body increases or decreases, or their separation distance varies, changes take place in their mutual force of attraction or gravity. Smaller planets exert lesser gravitational forces than larger ones of the same density because they are less massive. Gravity is used as a force in many natural energy systems, such as falling or moving water.

One very promising use of off peak electrical power involves the pumping of large volumes of water to an elevated storage reservoir for later release through turbine hydroelectric generators at peak demand times. Such a process may save the capital cost of oversized conventional and standby generators, especially when implemented at large scale.

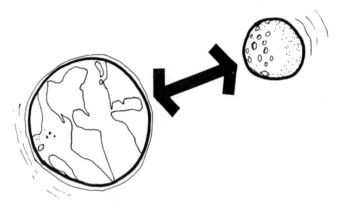

gravity is the attraction between two masses

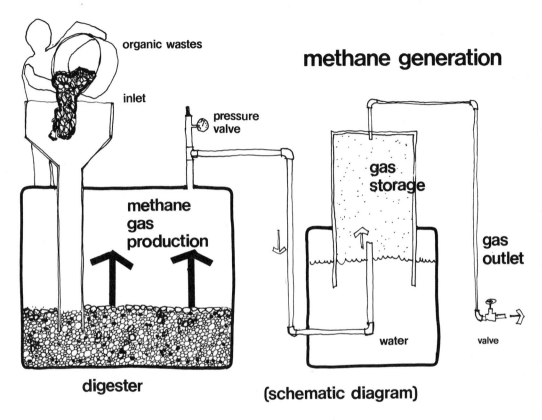

organic wastes

inlet

methane generation

pressure valve

methane gas production

gas storage

gas outlet

digester

water   **valve**

(schematic diagram)

Of course, geothermal and gravitational energy are only two of the types of energy provided by the earth. Others, including the fossil fuels, coal, oil, natural gas, and nuclear, have already been mentioned.

Methane composes about 95% of natural gas, and is found in oil wells and coal mines. However, it is also produced as a by-product of the decay of organic materials in the absence of oxygen. The presence of methane has been detected in swamps, septic tanks, landfills, and the digestive systems of animals.

In a landfill at Palos Verdes, near Los Angeles, thirteen million tons of refuse have accrued since 1957. This refuse is now generating methane, and one contractor hopes to be able to recover 28.32 cubic meters (1000 cubic feet) of gas per minute from it. If this project is a success, there are numerous other landfills around the country to which the same techniques could be applied.

Methane can also be derived from a digester, into which organic materials are fed. This type of digestive process takes place in several stages. First, organic matter is mixed with water to form a slurry, which is loaded into the digester tank. Aerobic bacteria – those requiring oxygen – begin to break down the material into water and carbon dioxide. When the oxygen inside the tank is used up the aerobic bacteria die, and anaerobic bacteria go to work. Without the benefit of oxygen, these bacteria are not able to break down the organic matter as completely as the aerobic bacteria. Methane is one of the products of this anaerobic fermentation and is collected for use. The uses of methane are as varied as the uses of natural gas, but methane produced in a digester is a renewable fuel and has applications for transportation, synthetic materials, and heating.

notes

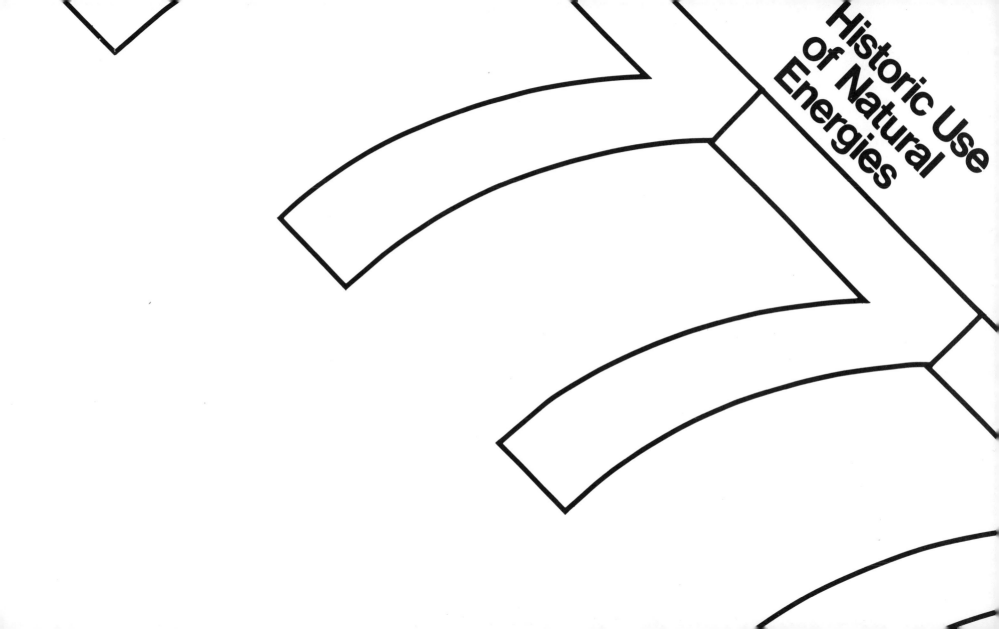

Historic Use
of Natural
Energies

Throughout history, at least until the Dark Ages in medieval Europe, man has allowed his architecture to evolve in harmony with natural climatic conditions. The sun and the natural forces which are available to him have been utilized in various ways. Changes in his technology and culture have fostered changes in his attitudes concerning his utilization of nature's energy. Characteristics of available materials have also had an impact on the architectural forms predominantly used by various civilizations. Some of the numerous ways in which these natural forces have been used are presented in this chapter.

The energy of the sun in combination with the earth provides the food energy used by man and all other living creatures. Plants are the first major link in all food chains, as well as the receptors of the solar energy which is eventually stored as fuel. Many plant fibers are also used as the construction materials (wood, straw, etc.) which are the basic units of shelter. Other materials for shelter (steel, concrete, brick, stone, mud, etc.) are derived directly from the earth or are produced from natural materials which have been refined in some way.

Primitive man in temperate zones used the earth and its topography as he found it, to provide himself with shelter from the environment and protection from wild animals and enemies. He used caves and the undersides of cliffs for his habitat, and showed little desire to fabricate his own dwellings. His attitude toward his shelter produced the least possible negative impact upon the environment. These natural earth dwellings provided micro-climates in which to live, less severe than the outside climate. The thermal inertia and insulation of the earth produced a cool home in the summer and a warm home in the winter.

As man's needs developed, he started constructing his shelter using the materials he found around his encampment. He learned that entries and windows facing the sun and solid walls intercepting the harsh winter winds were the most comfortable configuration. This orientation was prevalent until the advent of industrial man, who, through his advanced technology, decided he could provide a comfortable interior environment by expending fossil fuels no matter how he oriented doors, windows, etc. Topography and the availability of resources played a dominant role in dictating how early buildings and communities would be situated. Land forms were used to separate classes within societies, as well as to provide protection from climatic forces and hostile neighboring tribes.

primitive man used the natural energies and materials around him

Egyptian designers understood how to use natural climatic forces to provide comfort in their buildings. They used the sun to warm space and to provide interior illumination. Thick walls, warmed by the sun, were used to create uniform nighttime temperatures; in turn, the stored heat was used to induce natural air movement which ventilated buildings. The walls were effective even after the sun had set because of the heat they contained.

The Egyptians were the first known civilization to produce a sun-dried brick. Walls 20 to 30 cm (8 to 12 inches) thick, constructed out of these bricks, could develop a thermal time lag of 6 to 10 hours.

Interior courtyards and pools of water were used in structures built by the Aegean civilization to provide natural lighting to interior and exterior spaces, as well as to control temperatures in these areas. The design of entryways and court areas encouraged the natural flow of air through interior spaces. Runoff water was collected in the pools located in interior courts. The water provided evaporative cooling and humidification for the building.

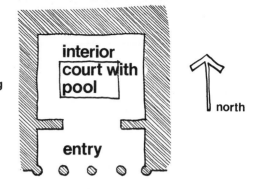

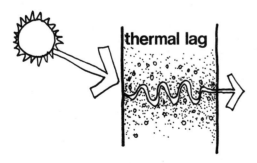

The Greeks were worshipers of the sun, and this devotion is amplified in their architecture. Their public meeting places, agoras, were squares situated so that they would receive the sun's natural warmth. They were surrounded on three sides by buildings with the fourth side open to the south sun. Private residences were arranged so that each had a prominent southern exposure. Hippocrates in his "Air Light Treatise" declared that cities should be sited so that buildings and streets would be penetrated by the sun and wind for hygienic purposes and odor removal. (Many streets also served as open sewers.) He suggested also that cities should be located where pure water was available, and that marshlands should be avoided. The Greeks realized and exploited the fact that the sun, used to cast shadows, would bring out the detail and form of buildings.

The Romans were located at much the same latitude as the Greeks and were equally fascinated with the sun. Vitruvius, a Roman builder and architect-philosopher, wrote the "Ten Books of Architecture" in which he presented many energy conserving ideas relating to siting, orientation, and climatic response. "In the north, houses should be entirely roofed over and sheltered as much as possible, not in the open, though having a warm exposure. But on the other hand, where the force of the sun is great in the southern countries that suffer from heat, houses must be built more in the open and with northern or northeastern exposure. Thus we amend by art what nature, if left to herself, would mar. In other situations too, we must make modifications to correspond to the position of the heaven and its effect on climate."

Vitruvius incorporated many passive energy systems into his designs. He was greatly concerned that his planning and layout of cities and buildings would allow the maximum use of natural energy forces. He included many concepts in his designs which had been presented by earlier writers; most of these earlier works have since been lost. On the use of sun and natural light he wrote, "There will also be natural propriety in using an eastern light for bedrooms and libraries, a western light in the winter for baths and winter apartments, and a northern light for picture galleries and other places in which a steady light is needed; for that quarter of the sky grows neither light nor dark with a course of the sun, but remains steady and unshifting all day long." The need for climate responsive architecture is reflected in his statement that, "One style of house seems appropriate to be built in Egypt, another in Spain, a different kind in Pontus, one still different in Rome, and so on with lands and countries of other characteristics."

The observations of Vitruvius have remained valid, though often ignored, through history. To be in harmony with his natural surroundings, man's architectural forms must respond to the influences of regional climates. In all areas, dwellings should respond to the natural forces of the sun and weather. They should harmonize with the environment and employ as many local building materials as possible.

winter igloo

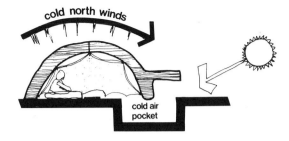

cold north winds

cold air pocket

summer hut

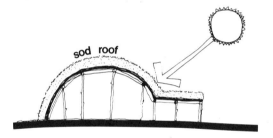

sod roof

In cold, humid northern climates, dwellings must be constructed so that heat loss is minimized and heat gain maximized in winter. In sparsely populated Arctic regions the igloo is the most common type of building. An igloo is built in the shape of a dome which gives it a number of advantages. The dome shape encloses the maximum amount of internal volume within the least external surface area, an important consideration in controlling heat loss.

To minimize the heat lost to prevailing north winds, and to realize a small amount of solar heat gain, the only opening is on the south side. The opening is actually a tunnel which dips below grade, then back up into the interior space. Since cold air does not rise, it settles in the lowest spot in the tunnel and does not gain entry into the interior. An igloo is constructed of ice and snow, local materials, which form a vapor barrier to the outside. Animal skins are stretched across the walls and ceiling areas on the inside; these create dead air spaces between the skin and the ice, providing insulation and restricting heat loss. The majority of the heat in such a space is provided by the occupants, and the average indoor temperature is maintained at 4.4 C (40 F). The Eskimo's comfort range is considerably lower than that of the average American, and because of their warm clothing, they feel no discomfort at this low temperature. During extremely cold periods a fire can be built inside the igloo, the smoke being vented through a wooden exhaust stack which penetrates the top of the igloo.

Summer huts are constructed in a manner similar to the igloo, but of different materials. Sod is used on the roof to act as insulation and also to seal the surface to protect against moisture leaks. These huts are oriented with the back facing the north for protection from the cold night winds, and the entry facing the sun for maximum solar heat gain.

In wet, humid climates, cooling is the primary consideration. Buildings are left as open as possible to allow the maximum amount of air movement from the exterior to the interior. The area of the roof is minimized to decrease solar heat gains through it. Usually roofs are constructed of materials which are light colored so that heat is reflected from them. A grass or thatched roof will allow air movement through it but will not allow water leakage. Frequently, penetrations are provided in roofs to permit ventilation of upward rising warm air. In regions with high precipitation, buildings are raised above the ground to prevent damage resulting from water runoff. All the materials used in the construction of these shelters have low thermal mass and high thermal transmission values to prevent heat buildup during the day.

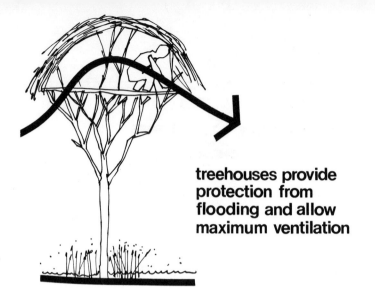

treehouses provide protection from flooding and allow maximum ventilation

buildings on stilts provide space in water areas

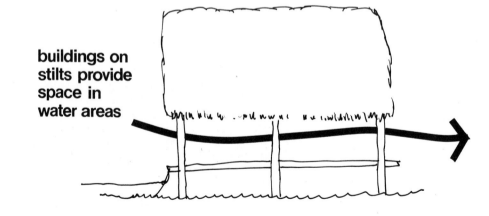

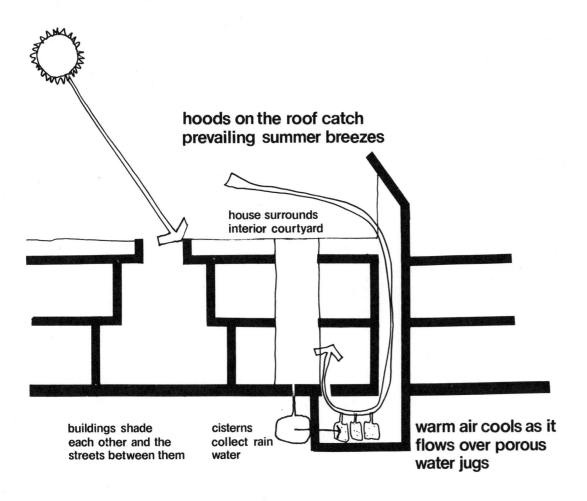

hoods on the roof catch prevailing summer breezes

house surrounds
interior courtyard

buildings shade
each other and the
streets between them

cisterns
collect rain
water

warm air cools as it flows over porous water jugs

In hot, dry climates, cooling and humidification of the air are necessary. These can be accomplished using the maximum possible air movement and evaporative cooling. In parts of India and Africa, many buildings are close together, restricting potential for cross-ventilation, so other means of inducing air movement are employed. Air scoops which catch the prevailing winds and direct them downward into the building are attached to roofs. The air travels down a large duct (approximately one square meter) to the basement or cellar area in which jugs filled with water are kept. The jugs are made from a porous clay to allow for maximum evaporative surface. As the air passes over the water, it is cooled and picks up moisture. The conditioned air is then directed upward through the building by warmed air leaving the stack above.

Most houses are situated around an interior court which acts to promote air circulation and collects any precipitation. Since most buildings are very close together, very little heat from the sun is gained through exterior walls. Roof surfaces are light colored, reflecting heat upward.

A hot dry climate is also prevalent throughout the southwest region of the USA. In this area the American Indians built many energy efficient community developments. One of these has the Spanish name Mesa Verde, and is located in Colorado. In this area the sandstone was eroded by wind and water, forming recesses below the edges of cliffs. These recesses were cave-like and provided shelter and a natural defense barrier for the earliest people in the area. As the culture of the region developed, dwellings were built in the backs of the caves, using the local materials of adobe and short lengths of timber. The overhanging cliffs not only provided natural defense from enemies but also offered protection from the intense summer sun. The design of the community provided for collection of solar radiation during the winter. Because the winter sun is low in the sky, buildings received sun during the day. The heat was stored in the massive adobe walls then reradiated to the cave area at night. The cave and the buildings contained in it had large surface areas for heat gain. The development is sheltered from cold northern winds, but is open to the prevailing summer breezes. The land above the cliffs was used for farming, hence the name Mesa Verde (green table). Because the people of this area had protection from harsh climatic elements and used natural energies to provide their comfort, they were able to enjoy a higher standard of living than other tribes which lived in this region.

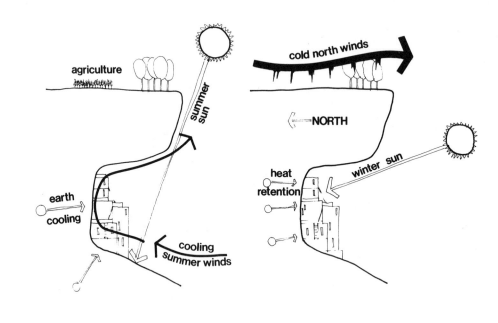

summer

winter

mesa verde

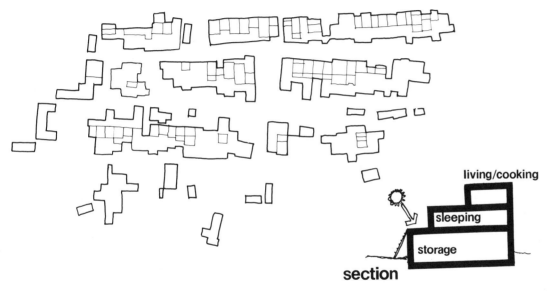

living/cooking

sleeping

storage

section

site plan

acoma pueblo

South of Mesa Verde in New Mexico was the Acoma pueblo. The Indians who built this complex placed it on top of an inaccessible mesa, so that it could be easily defended. It had many design features which modern man should incorporate into his buildings. The entire complex was tiered which allowed the maximum surface area to be exposed to the sun. Each level had a specific function which promoted the maximum utilization of natural energy. The lowest level had very few openings and was used for storage.

Since heat rises this was also the coolest area. The roof over the storage space was used as a terrace and was accessible from the second level, which served as a sleeping area. The upper story was used for living, cooking, and eating. Because of the stacking of all these levels, natural ventilation was supplied to the upper level to exhaust heat and odors. The pueblo had numerous houses tiered in this way and connected side by side. With only two walls exposed to the outside, heat loss was reduced. The dividing walls between units were several feet higher than the roof. This provided privacy, but more importantly provided shade for the terrace work areas. The walls were constructed of adobe and reinforced with timber, and consequently were high in thermal inertia which moderated inside temperature year round. As the sun would strike the walls, the heat would generate convective air currents along them, which added to natural ventilation. The roofs of the units were made of lightweight materials offering low thermal inertia and high heat transmission; because of them, summer heat would not accumulate in the buildings.

Another Indian development in New Mexico was discovered by western man in 1921: Pueblo Bonito in
Chaco Canyon. It is similar in design and construction to the Acoma pueblo, having three levels, but
was further unified by a curvilinear plan oriented to
the south. High walls on the sides of the pueblo controlled the time of day at which the sun entered and
exited interior areas. This pueblo was built in two
stages; remarkably, the addition of the second phase
further increased the overall energy efficiency of the
complex. Similar materials were used in the Acoma
and Bonito pueblos, which maximized the utilization
of energy in each.

pueblo
bonito

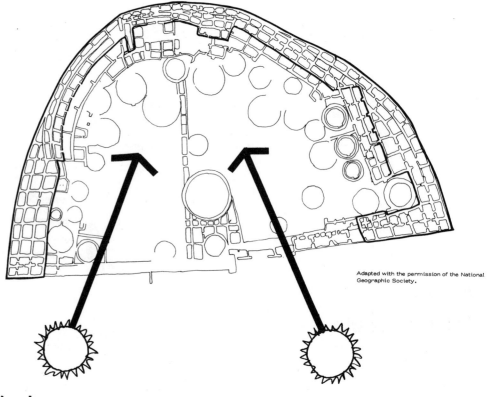

Adapted with the permission of the National
Geographic Society.

site plan

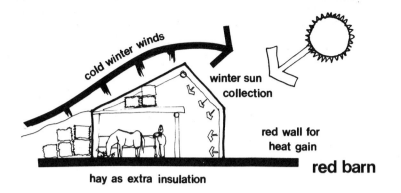

cold winter winds

winter sun
collection

red wall for
heat gain

red barn

hay as extra insulation

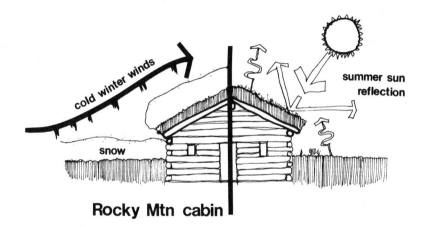

cold winter winds

summer sun
reflection

snow

Rocky Mtn cabin

The Indians were not the only Americans who built their structures to be responsive to climatic forces. Barns in many parts of the country were built with a long sloping roof facing into the direction of prevailing winter winds, and terminating several feet above the

ground. The space between the ground and the roof was filled with hay, straw, or other fibrous grasses to provide an insulating barrier. Snow covered the grass, which added to its insulating value. The south wall was painted red ("paint" was made by combining skim milk, iron oxide, and lime) to absorb the solar radiation striking it. With the protection from winter winds and insulation on the north, the absorption of solar energy on the south, plus the body heat of the animals on the inside, the barn functioned as a passive solar collector, providing warmth and protection to the farmyard occupants. This roof form also distinguished the numerous New England "saltbox" houses of the same era.

Early settlers in the Rocky Mountains used the materials they found there to build cabins which provided them with year round comfort. Structures were set into the ground at least two to three feet below grade to take advantage of the stable ground temperatures. Large logs were used to construct the walls and roof which provided insulation, thermal inertia, and heat time lag. In addition to the logs, the roof was covered with a layer of sod, forming an insulating barrier which minimized the heat loss and gain through it. In the winter, snow would accumulate to the height of the lower edge of the roof, producing a streamlined shape to minimize the heat lost to winter winds. The furry animal skins which covered the windows, plus the snow depth, greatly decreased heat loss through wall and window areas. In the summer, the skins were removed and the cooling breezes blew freely through the interior. The roof reflected most of the day's heat, while ground temperatures helped maintain comfort.

Many other examples can be cited in which people throughout history, faced with adverse climatic conditions, adapted their architecture to high energy efficiency. Many ideas found in primitive architecture can be incorporated into modern designs to promote harmony with nature instead of dominance over it. To waste our energy and that stored in our limited resources simply to overcome nature is a futile undertaking.

Most of the stated examples show how solar and other natural energies have been used in a passive way. Historic examples of active systems are also available.

One of the first recorded uses of solar energy was made by the Greek mathematician and engineer Archimedes. When the Roman fleet was attacking Syracuse harbor in 212 B.C., he allegedly set the ships on fire by focusing the rays of the sun onto their sails using approximately 4000 gold and bronze shields. It is reported that the sails were black, which helped to make this amazing feat possible.

A twelfth-century writer has reported that a similar solar display was mounted at Constantinople in 600 A.D. when a mirror focusing system designed by Proclus was used to ignite ships in the fleet of Vitellius. Other pyrogenic military applications have been reported through history.

One of the first machines which could harness the sun's energy and convert it to mechanical work was invented by a Frenchman, Solomon de Caux, in 1615. In his machine, air was expanded by the sun's heat. This expansion caused an increase in internal pressure which was then used to pump water.

Another Frenchman, Antoine Lavoisier (1743-1794), used a 128 cm (51 inch) diameter lens to concentrate the sun's rays and develop temperatures of up to 1755 C (3190 F) which were sufficient to melt metals. His lens was filled with white wine because it had better optical properties than the glass that was available to him. This was the predecessor of today's solar furnaces.

In the 1870's another Frenchman, August Mouchot, developed one of the first solar powered steam engines, which was used to drive water pumps; the steam was used to produce ice by an ammonia water absorption process (1878).

In 1859 oil was discovered at Titusville, Pennsylvania. This heralded the beginning of an area in which oil and gas would be readily available at low prices. Because fossil fuels were readily obtainable, many people lost interest in using the sun to power machinery, except in the most remote desert areas where transportation was difficult.

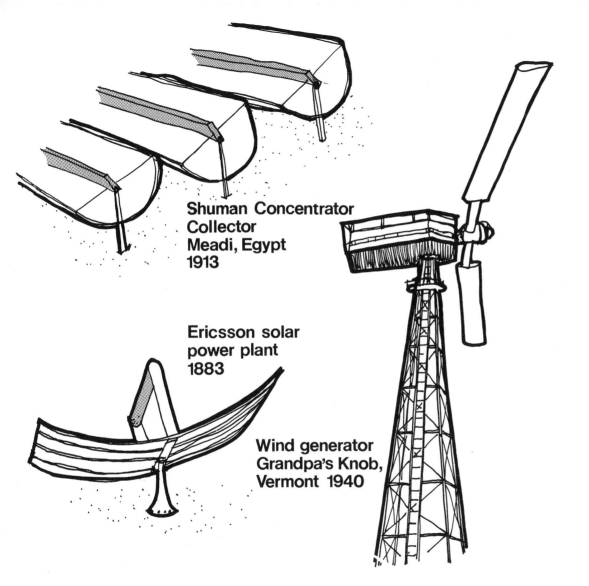

Shuman Concentrator Collector Meadi, Egypt 1913

Ericsson solar power plant 1883

Wind generator Grandpa's Knob, Vermont 1940

Isolated examples of solar design appeared in the 1870's, when John Ericsson, an American inventor, designed and built several solar powered hot-air engines which adapted to conventional steam power for cloudy day operation. Even though these were the most effective such engines up to that time, he admitted that they could not compete with the low initial cost and high reliability of steam power generated by burning coal.

In 1901 A. G. Eneas built a large solar reflector which was used to power water pumps on an ostrich farm in Pasadena, California. The pumping system was rated at four horsepower and could pump 1,400 gallons per minute against a static head of 12 feet.

In 1908 Frank Shuman, another American, established one of the first solar flat plate collector companies, Sun Power Company, which in 1911 built a demonstration plant of more than 10,000 square feet in collector area. The system operated a steam driven water pump. In 1913 the company supplied an installation with 13,000 square feet of parabolic trough concentrator area in Meadi, Egypt (south of Cairo) which powered a 100 hp steam engine.

In the early 1940's a large scale wind driven electrical generator was installed at Grandpa's Knob, Vermont. The machine was conceived by Palmer C. Putnam, an engineer, who coordinated the design and construction of the 1,250 kilowatt machine. It had a useful life of only 18 months, eventually losing a rotor blade which was never repaired because cost calculations indicated that it could not compete economically with generators powered by fossil fuels.

By the 1930's small wind powered electrical generators were gaining popular acceptance with people living in remote areas of the plains and western USA. During the Roosevelt Administration of the 1930's, however, the Rural Electrification Administration (REA) was established which provided federally subsidized power to these areas and privately owned wind generators soon lost their appeal.

Only recently has universal interest in solar, wind, and other natural energy been revitalized. The high costs of power and heat supplied by public utilities, and the shortages of fluid fossil fuels have forced many people to reevaluate their long term energy needs and the means by which they will be supplied. The necessity for climate responsive architecture and systems has reemerged.

Contemporary
Use of Energy

Some buildings in the United States lose as much as 50-60 Btu of heat per hour per square foot of floor area (158-189 watts/square meter) during the winter. On a yearly basis buildings directly consume 15% of all the energy used by this nation. If the energy required by buildings could be reduced by 80% through energy conservation and by using alternative energy sources, a decrease in the nation's total energy usage of 12% could be realized.

The American fascination for detached single-family homes has produced numerous problems, the majority of which have perpetuated excessive energy usage. Each member of a family must use an automobile to travel to and from his single-family suburban home. The population density in most suburban communities is not sufficient to develop economical mass transit or other systems for moving people. The transportation of goods and materials to suburban locations promotes further energy usage.

A major portion of the energy utilized for space heating is consumed by single-family houses. Consumption by larger buildings is controlled by many conditions which can be modified on existing structures and incorporated into the design of future buildings. The areas where the average house loses heat are illustrated on the accompanying diagrams. Each of these areas can be treated and improved separately to diminish winter heat losses. Many of these and other characteristics can also be analyzed to decrease summer heat gains.

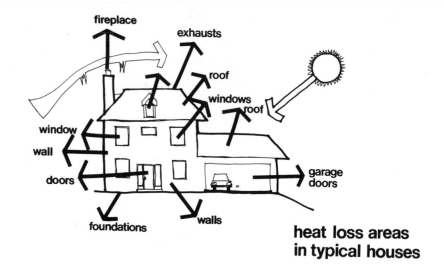

heat loss areas in typical houses

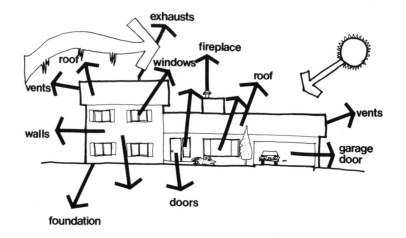

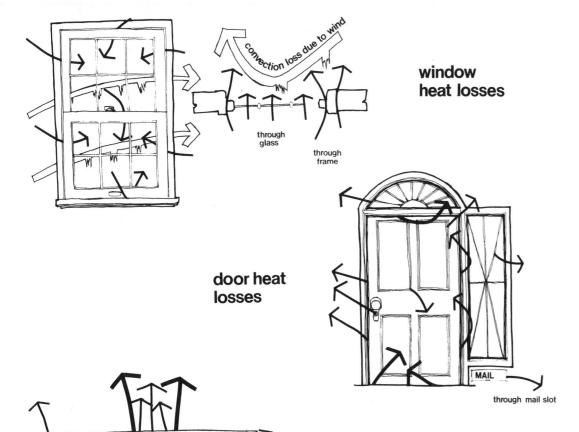

**window
heat losses**

convection loss due to wind

through glass

through frame

**door heat
losses**

MAIL

through mail slot

fireplace heat losses

A great deal of heat is transferred through doors. It flows through cracks in the wood of the door, around its edges, through keyholes, transoms, mail slots, adjacent windows, and is conducted through the door itself.

Heat is lost through windows anywhere there is a crack in the frame or the glass. Losses through the glass itself can be excessive. Heat is conveyed to the inside surface of the glass by convection and radiation, conducted through the glass, then lost to the outside by convection and radiation.

Fireplaces are traditionally identified with warmth and comfort. The negative aspects of continual interior heat losses through the flue and chimney which project through the outer wall or roof, and the depletion of interior space oxygen as a result of the direct combustion process during cold periods are generally not considered. A large percentage of the heat produced by a conventional fireplace goes up the chimney. Fireplaces also contribute to the heat loss of a house by drawing warm air out of the room through the flue after a flow has been initiated. Dampers on most units leak, allowing heated air to flow out of the room, even when the fireplace is not in use. Whenever air is removed from a room, it is replaced by cold air which infiltrates from the outside. The net effect of most fireplaces is cooling of a house rather than heating it.

There are a number of ways that a conventional fireplace can be modified to overcome these energy losing aspects and improve their efficiency. To understand why some of them are effective, it is necessary to analyze what happens to the heat liberated in a fireplace.

115

Much of the heat is dissipated up the chimney while some is radiated to the floor, back, and sides of the firebox and the remainder is radiated to the house. Heat radiated to the sides and back will be conducted to the outside if adequate insulation is not provided. Insulation between the firebox and the outdoor projection of the chimney or a chimney and firebox built within the insulated outer walls will increase the quantity of heat radiated and convected to surrounding interior areas. If the walls and floor are masonry, they will store the heat and gradually release it to the house.

In this regard, the advantages of interior plan location of fireplaces are quite pronounced, as their lateral heat losses are confined within the building. Very tight or airlocked double dampers are needed to avoid continual losses through the flue. During the winter, they should be closed at all times, except when the fireplace is in use. During the summer, a properly designed flue may be used as a vertical stack to initiate passive ventilation. Direct outdoor air for combustion, brought in through a duct that is in use only when a fire is present, conserves energy and helps to maintain interior oxygen levels.

Many devices and prefabricated firebox units are available which employ convection to more efficiently distribute the heat from a fireplace. The prefabricated fireboxes have low level cool air intake slots in the front or on the sides and warm air outlets on the top. The air is circulated around the back and sides of the firebox, where it is warmed and then returned to the room, or ducted to adjoining areas. A number of manufacturers are marketing grates

made from hollow metal pipes which are heated by radiation and convection. Air blown through the pipes absorbs heat before re-entering the room, delivering heat which otherwise would have been lost.

Control over the supply of oxygen for clean but not too rapid burning of the fuel should be considered in the choice of fireplace or wood burning stove details. Heat tempered glass firescreens for fireplaces, and Franklin stoves with draft control, provide this function. Glass firescreens also afford safety by preventing sparks from being thrown into a room. Heat is radiated and convected to the glass, conducted through it, and then radiated and convected to the room. These glass firescreens help to prevent heat loss when the fireplace is not in use.

Devices are also available which attach to the stack or chimney and extract heat from the exhaust air which can be routed back to the home. Some of these are heat exchangers of the air to air variety, while others contain heat pipes. The efficiencies of these devices vary from one type to another; in fact, one confident manufacturer warns against cooling the exhaust gases too much or the chimney will no longer draw.

A well designed fireplace, arranged as a central feature of a home or building, may be combined with an insulated chamber that would also retain either solar thermal or off-peak electric energy. This heat could be delivered to the building as necessary to maintain comfort.

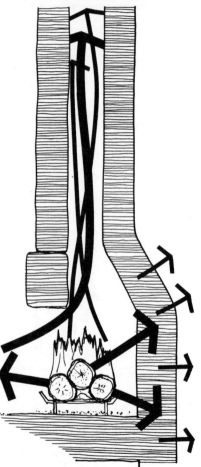

fireplace loses heat through back and out the chimney

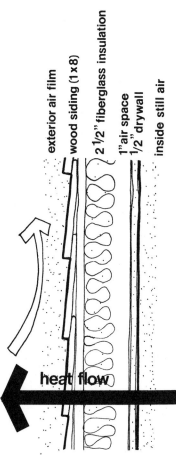

exterior air film
wood siding (1x8)
2½" fiberglass insulation
1"air space
½" drywall
inside still air

heat flow

wall cross-section

Many factors affect the heating and cooling "loads" of a building, which can be thought of as the amount of heat that must be added in the winter and taken out in the summer to maintain the comfort of the occupants. Some of these factors are the size and shape of the building, the occupancy and tasks performed there, climatic conditions, orientation of the building, the layout of the rooms, and the rate at which heat enters or leaves the building. The materials selected for construction and the thickness and type of insulation used control the rate at which heat will flow through the walls, ceilings, roof, and floor of a building. This heat flow rate can be calculated in terms of Btu per square foot per hour per degree Fahrenheit temperature difference between the inside and outside surfaces (Btu/sq. ft./hr./deg. F). To make this calculation, specific construction details must be known, and certain properties of the construction materials must be determined. For specific information about thermal properties of materials the ASHRAE Handbook (American Society of Heating, Refrigeration and Air Conditioning Engineers), 1972 Edition, is the best reference.

Following are sample calculations to determine the heat flow rate through the walls, roof, floor, and openings of a typical suburban home. Some assumptions about the temperature differential are made, and the heat loss in terms of Btu per hour (Btuh) is determined. Suggestions are made for modifying the house to make it energy conserving, and the heat loss for the modified home is calculated.

The construction of the wall and the material inside the wall must first be determined. All the materials have heat conductivity and heat resistance values. One is the inverse of the other. Resistance factors are easiest to work with since they can be added directly to determine the resistance of a complete section. The larger the resistance factor (R factor) of a material, the better insulating qualities it will have. The films of air on the exterior and interior surfaces of a wall also act as insulators and are assigned an R factor.

Below are the R factors for the individual components of a typical suburban house wall with little insulation. Also given is the total R factor of the wall. The values are from the ASHRAE Handbook, 1972 Edition.

	R Factor
Exterior air film	0.17
Wood siding (1x8)	.78
1" air space	.90
2½" fiberglass insulation	7.80
½" drywall	.45
Inside still air	.68
Total of R Factors, R_t	10.78

The rate of heat transfer through the wall is expressed by the overall coefficient U or U-value, which is expressed in units of Btu per hour per square foot per degree Fahrenheit (Btu/hr. sq. ft. deg. F). The U-value is the reciprocal of the total R factor, $U = 1/R_t$. For this wall the U-value is $1/10.78 = .093$. The smaller the U-value the better the insulating properties of the wall. For the wall section above, if 1" of rigid polyurethane foam insulation (R = 5.88) were added in place of the air space, the total R factor would be 15.76, resulting in a U-value of 1/15.76 = .063. Changes could also be made to the siding, interior drywall, and other materials which would decrease the U-value.

To calculate the amount of heat flow per hour through the wall the U-value must be multiplied by the wall area and the temperature differential, that is, the difference between the outside and inside air temperatures. In Denver the design temperatures are -10 F outside and 70 F inside. This is an 80 degree differential. If the wall mentioned above, with a U-value of .093, had an area of 900 square feet (excluding doors and windows), the heat flow per hour could be calculated as:

Wall area x U-value x temp. diff.   = Heat flow
   900 sf   x   .093   x   80 F   = 6696 Btuh

This type of calculation assumes that the temperature differential is constant throughout the wall height. It also assumes "steady state" heat conditions, as well as a homogeneous wall section. Strictly speaking, adjustments should be made to the heat flow for areas of the wall which are occupied by structural, electrical, and mechanical elements. In some cases, this could be in excess of 30% of the wall. The U-value in these areas would be greater than the one calculated, meaning more heat flow through the wall. These considerations also apply to the roof area, and structural ground floors.

The heat flow through a wall with the extra insulation is decreased to 900 x .063 x 80 = 4536 Btuh. With the extra insulation, the wall heat flow per hour is calculated as 900 x .063 x 80 = 4536 Btuh, a decrease of 2160 Btuh.

The heat flow through the ceiling can be calculated in a similar manner. If the attic is vented to the outside, the calculations for a typical ceiling are presented below.

Ceiling	R Factors
Outside air	0.17
Insulation (3")	8.33
½" drywall	.45
Inside still air	.61
Total of R Factors	9.56

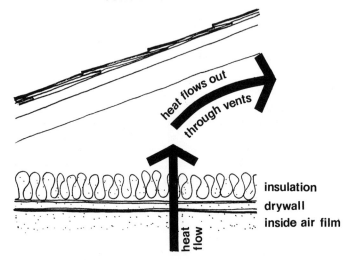

heat flows out through vents

heat flow

insulation
drywall
inside air film

The U-value is 1/9.56 = .105 Btu/hr./sq. ft./deg. F.

If the vents in the attic are sealed, the ceiling, attic, and roof contribute to the insulation value. With additional insulation and the attic vents sealed, the calculations become:

Ceiling, Attic, and Roof	R Factors
Outside air	0.17
Shingles (Asbestos-cement)	.21
Plywood ½"	.62
Attic air space	1.40
6" fiberglass batts	18.72
Existing insulation	8.33
½" drywall	.45
Inside still air	.61
Total of R Factors	30.51

The U-value is 1/30.51 = .033 Btu/hr. sq. ft. deg. F.

Assuming an approximate projected roof area of 24' x 32' = 768 square feet, where 24' and 32' are the plan dimensions of a rectangular building, the maximum heat flow per hour is 768 x .105 x 80 = 6451 Btuh. If the alterations are made to the roof, the heat flow per hour is 768 x .033 x 80 = 2027 Btuh. This is a significant reduction. The projected roof area corresponds closely to the ceiling area of the house, and is used instead of the actual area of the roof, because the heat is assumed to be out of the house once it crosses the plane of the ceiling.

Depending on their construction and location in relationship to the grade, heat can be lost through floors. In this example it can be assumed that the house is atop a basement. The total floor area of the basement is 750 square feet and the total basement wall area, exposed to the ground outside, is 800 square feet. For a ground water temperature of 50 F and a basement temperature of 70 F, the hourly basement floor loss is 2.0 Btu/sq. ft. and the hourly basement wall loss is 4.0 Btu/sq. ft.

The heat flow through the concrete is: 750 x 2.0 + 800 x 4.0 = 4700 Btuh.

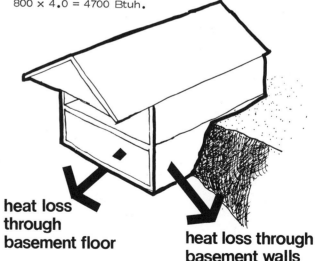

heat loss through basement floor

heat loss through basement walls

A great deal of heat can flow through the windows of a building. Assuming that the typical house has single pane non-insulating glass windows, their U-value is 1.13. If the window area is 20% of the building floor area, then the heat flow through the glass is: (768 × .20) × 1.13 × 80 = 13,885 Btuh. For a metal sash in which the glass area is 80% of the window area, the adjustment factor is 1.0, with a resulting heat flow of: 13,885 × 1.0 = 13,885 Btuh.

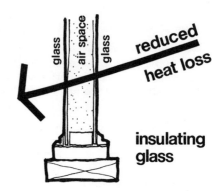

reduced heat loss

insulating glass

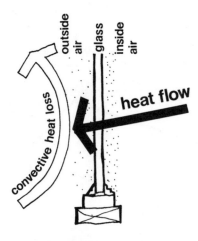

heat flow

If wood frame insulating storm windows are used, in which the glass area is 60% of the window area, the adjustment factor is .80 and the U-value is reduced to .46, resulting in a heat flow of 153 × .46 × 80 = 5630 Btuh.

heat loss through door

If insulating glass (2 panes with ½" air space between) is used, the U-value is .58. The heat flow through the glass can be calculated to be 153 × .58 × 80 = 7099 Btuh. For a metal sash, double glass, and glass area of 80% of window area, the adjustment factor is 1.20, so the heat flow is: 7099 × 1.20 = 8519 Btuh.

Heat is also lost through doors. In this example assume that a standard 1½" solid core door and a metal storm door are used. The U-value for the doors is .33. There are two doorways, front and rear, each 3' × 7'. The heat flow is (7x3)2 × .33 × 80 = 1109 Btuh.

By replacing the conventional doors with insulated doors, the U-value can be reduced to .15 with a heat flow of 42 × .15 × 80 = 504 Btuh.

Infiltration is the replacement of some or all of the air in a room or building by air from the outside. In the winter, cold winds blow outdoor air into interior spaces through cracks around windows and doors, as well as through other cracks in the building. The cold air forces out air that has been warmed by heating units. A great deal of heat is lost in this process. The procedure for calculating the amount of heat lost as a result of infiltration involves first determining the amount of air displaced per hour, then multiplying this by the amount of heat contained in a unit volume of air and by the design temperature differential.

Different rooms of a building experience different numbers of air changes per hour, depending on the number of exterior walls, doors, and windows. The ASHRAE Handbook contains lists for this purpose.

In this example we'll assume that 30% of the building is composed of rooms with doors or windows on three sides, and the remainder has doors or windows on two sides. So, 30% of the building has 2 air changes per hour and 70% has 1.5 air changes per hour. To determine the amount of air displaced per hour, the room volume is multiplied by the number of air changes. The total floor area of the building is 768 square feet and with 8' ceilings its volume is: 768 × 8 = 6144 cubic feet. With 30% having 2 changes per hour and 70% having 1.5 changes per hour, the total amount of air displaced is: (6144 × .30) × 2 + (6144 × .70) × 1.5 = 10137.6 cubic feet of air.

The heat capacity of air is calculated by multiplying the specific heat of air (Btu/lb./deg. F) by its density (lb./cu. ft.). Representative values for these are 0.24 Btu/lb./deg. F and 0.0624 lb./cu. ft. (sea level air density is 0.075 lb./cu. ft.), with a product of 0.015 Btu/cu. ft./deg. F. The total heat loss due to infiltration is: 10,137.6 × 0.015 × 80 = 12,165 Btuh. One of the largest heat losses of the entire building is due to infiltration! This is typical of most buildings. It is necessary, in energy conserving buildings, to decrease infiltration losses as much as possible. If weatherstripping is added to the windows and doors, the infiltration losses can be reduced by 1/3. This applies to all buildings, so for this example the heat losses are reduced to 8110 Btuh. Further reductions can be realized with double door airlock entries and by including a continuous vapor barrier in all walls and roofs.

A summary of all the heat losses is presented below along with the heat losses for the modified building.

	Existing	Modified
Windows	13,885	5,630
Infiltration	12,165	8,110
Walls	6,696	4,536
Roof	6,451	2,027
Floors (basement)	4,700	4,700
Doors	1,109	504
Total	45,006	25,507

By simple means of energy conservation, the original heat loss for the typical house was reduced by 43%. It is noteworthy that optimized interior planning, exterior reflecting devices, landscaping, and other measures would improve this figure considerably.

Similar methods for calculating heat flow apply to commercial and large residential structures. For many of these buildings the heat generated by lighting, internal equipment, and occupants may represent a large percentage of the total heat gain. In some cases these gains may be so large that heat has to be removed from the building, even during the winter. The occupancy of a building and the activities performed in it regulate the quantity of internal heat generated. For this reason these remarks are confined to characteristics of the building envelope and some mechanical systems which affect the heat gains and losses of large buildings and are directly related to energy conservation.

The areas where typical large buildings lose and gain heat are illustrated on the accompanying diagrams. Each of these areas can be modified and improved to increase the energy efficiency of existing buildings, and can be optimized during the design process of new buildings.

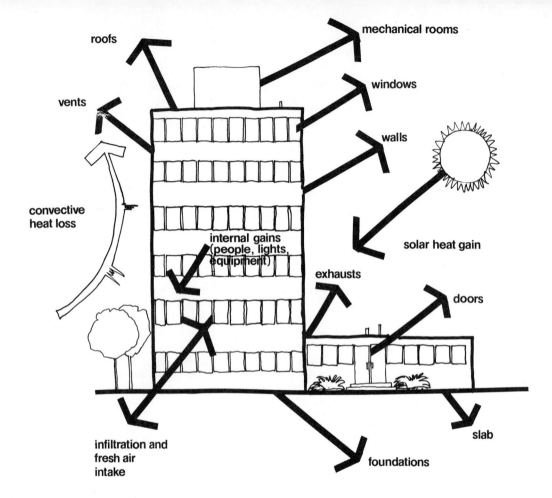

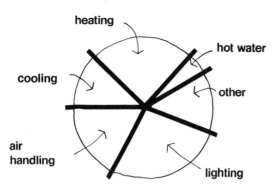

**annual energy consumption
of a school**

In the adjoining diagram the annual energy consumption of an average school in the northern United States is presented. This consumption is also representative of offices and other commercial buildings in this area, and clearly indicates that the largest percentage of the energy is used for lighting. For this reason natural lighting should be a primary consideration in building envelope design, yet heat gains and losses as a consequence of excessive window area should not be overlooked.

Large window areas contribute to winter heat losses and summer heat gains, resulting in year round expenditures of energy to maintain comfortable indoor environments. Glass is a notoriously poor insulator with single panes of flat glass having winter coefficients of heat transmission, U-value, on the order of 1.13 (R = 0.88). This coefficient can be reduced by about 40% if insulating glass is substituted for a single thickness of conventional glass.

An even greater reduction can be accomplished by reducing the glass area. A decrease in glass area does not necessarily mean an equivalent reduction in light. Depending on wall, ceiling, and floor color, as well as the placement of windows with respect to interior surfaces, a 50% reduction in glass area may result in less than a 30% reduction in light.

The design and integrity of the building envelope contributes directly to the infiltration of outside air into a building. The cost of heating and cooling this air can run as high as 50% of the total heating and cooling bill.

In a well sealed building, with weatherstripping, caulking, few cracks, and double "airlock" style doors, infiltration amounts to about one-half of the building's air volume change per hour. In a poorly sealed building the number of air changes attributable to infiltration may be five or six times this amount.

The insulating quality of walls, roof, and floors is among the factors contributing to the energy required for heating, cooling, and air handling. Adding insulation to these areas can be difficult and costly. However, by adding 2" of insulation to a brick cavity wall, the fuel required to offset heat losses and gains through the wall may be reduced by as much as 60%. Wall insulation, especially in multi-story buildings, can significantly decrease heat losses and gains, with a corresponding savings in the fuel necessary to maintain comfort.

Orientation of the building with respect to the sun and micro-climatic conditions has an effect on heating and cooling. Buildings with the majority of their windows oriented toward the south will receive beneficial winter sun but may receive too much summer sun. Properly designed shading devices can virtually eliminate summer radiative gains through such windows. On the other hand, north facing windows provide for summer cooling, but suffer tremendous amounts of winter heat loss. Further complicating this problem are buildings requiring cooling on the south to offset heat gains, and heating on the north because of heat losses, with no mechanical system to transfer internal heat from one area to the other.

All too often energy is consumed to cool one area of a building while additional energy is used to heat another area of the same building. Proper zoning of the mechanical systems and heat pump networks within the building can eliminate these wasteful practices.

Large classrooms, auditoriums, arenas, and other enclosed places of assembly can realize substantial heat gain even from seated occupants. It should be remembered that a sedentary human gives off approximately 400 Btuh to his surroundings. In a sports arena containing 20,000 spectators, hourly heat gain will easily exceed 8 million Btu, exclusive of energy expended on vocal support (cheering) or abuse (booing)!

In hot climates, cooling is of primary importance, and buildings with large surface area to volume ratios are best suited to these areas. In climates where heating is the primary consideration, surface area to volume ratios should be reduced. Considerable energy savings can be realized if ratios of this

type are computed and adapted to local climates. There is no practical way to alter these ratios for existing structures short of major reconstruction. Computations of this type are beneficial mainly in the design of new buildings or additions to old ones.

In conclusion, special attention must be directed toward identifying and quantifying internally generated heat. It may be the largest single energy input and is a factor to consider in the design of heating and cooling equipment. Building envelope characteristics regulate heat flow in a number of ways with the major concepts to consider being type, placement, and size of glass areas; quality, placement, and thickness of insulation; wall mass for thermal inertia and time lag; surface area to internal volume ratios; color and texture; as well as special shapes, wingwalls, overhangs, louvers, or shading devices. Building orientation for microclimatic conditions is essential. Prevailing winds steal heat and initiate pressure differentials which increase infiltration. Heat loss and gain due to infiltration must not be underestimated, for it can be substantial in a "leaky" building. Ventilation may require exhausting large amounts of heat to the outside. For winter operation, heat exchanges should be considered. Mechanical systems make the building "work" and should be designed to selectively allocate internal energy before drawing upon a supplementary energy source.

Building
Design for
Energy
Optimization

Lifestyles, as well as individual requirements, regulate the amount of energy consumed by a society. Some lifestyles are highly energy demanding while in comparison others are not. The recent trend in the USA has been to ignore energy requirements and to plan communities which, in their transportation and heating requirements, are extremely wasteful. The amount of energy required to maintain and operate new developments has been a secondary consideration in making decisions affecting land planning.

The environmental impact of new developments has, until recently, received little attention. Business, industry, and small municipal governments have encouraged urban sprawl. Shopping centers have been built in outlying areas which help to encourage growth of new communities. Industries have supported sprawl because it promotes the placement of value on certain items, such as the automobile. Municipalities, outlying large urban areas, benefit from growth because new residents mean a stronger tax base. Urban sprawl has also resulted in alarming increases in energy demand. All facets of delivering goods and services to people in divergent locations require energy.

Because the length of power transmission lines must be increased to reach suburban locations, energy losses in them are increased. Commuting from the suburbs to jobs in the city requires tremendous amounts of energy and personal time.

With each incremental increase in urban land area more material and energy is used for buildings, utility supply lines, sewer lines, roads, and other distribution systems. Because people have tended to move away from the city core, additional freeways and roads are needed. More cars and trucks are used to move people and goods from one point to another. With services and buildings spread over larger land areas, the necessity for using automobiles increases and so does the consumption of energy and the degradation of the environment.

Many urban problems have been caused by the exodus of the urban dweller to the suburbs. The commuter derives his income in one area and spends the bulk of it in another, thus depriving the city of sales tax and job opportunities. A city loses property tax money when someone moves out of it to the suburbs, or when buildings are torn down and parking lots are constructed to accommodate commuter's cars. People move from the city to escape air pollution, then contribute to it daily when they commute to their urban jobs. In cities such as Denver, Phoenix, and Los Angeles, where low density housing is predominant, there is a nearly exclusive reliance upon the automobile for commuting to work, shopping, and entertainment, with a resulting high ratio of automobiles to people. The air quality in these areas, meanwhile, has deterioriated significantly because of automobiles. The suburban commuter costs the inner city a great deal of money and contributes to its degradation by using city streets and services and not paying for them in his suburban taxes. He provides the city with its air pollution, parking problems, traffic control problems, freeway traffic jams, yet doesn't supply the tax money to solve them.

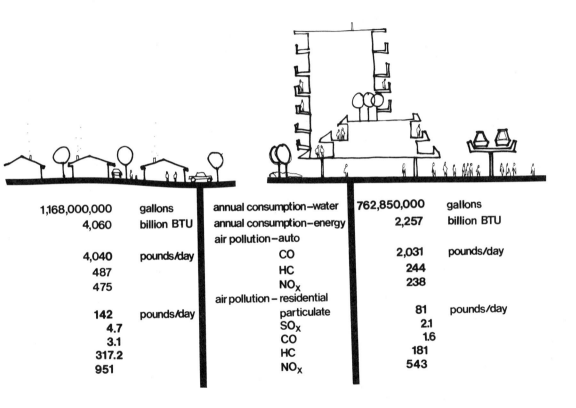

low density

high density

1,168,000,000	gallons	annual consumption–water	762,850,000	gallons
4,060	billion BTU	annual consumption–energy	2,257	billion BTU
		air pollution–auto		
4,040	pounds/day	CO	2,031	pounds/day
487		HC	244	
475		NO$_x$	238	
		air pollution – residential		
142	pounds/day	particulate	81	pounds/day
4.7		SO$_x$	2.1	
3.1		CO	1.6	
317.2		HC	181	
951		NO$_x$	543	

The closer people are to their work the less they need their automobiles. This may be one major justification for mass transit, which shows increases both in cost effectiveness and in logistic efficiency proportionate to population density. Few US cities have yet discouraged the use of automobiles in the core city areas. By contrast, many European cities have encouraged pedestrian and bicycle traffic by closing off inner city areas to most motorized traffic. This latter trend has now reached the more progressive of American cities; often such downtown concourses enjoy the additional amenity of special bus transport in restricted lanes.

Energy and water usage are lower in densely populated areas than in suburbs. The accompanying illustration compares the usage of these resources and the amount of certain pollutants produced by a low-density community and a high-density community. Each has 10,000 occupants. The land area required for the high density community is significantly less than the amount required for the low density one.

Individual buildings and building complexes should be designed to implement maximum energy usage in harmony with space use. One problem is the Western custom of designating for each room a single purpose. Other cultures are not found to allocate nearly as much building interior space per occupant, as do Americans. The number of rooms needed in a building decreases roughly as the multiple uses for each room increases. Many European families live in homes with less than 1000 square feet while most American families require 1500 –

2000 square feet of area. Much of the additional space is poorly designed and utilized. By employing multi-use of space, the area required for buildings can be reduced.

Energy and resource consumption is minimized by urban buildings, arranged in clusters and provided with multi-use rooms within individual living units and also provided with "common" spaces to accommodate activities for which the living units are not adequate. Buildings on the perimeter of the cluster will buffer ones on the inside from extremes of climate, especially wind, and should be designed for the more severe conditions. Optimized energy conservation is facilitated by the cluster and multi-use concepts.

Because the total floor area needed for functional individual living units is decreased by multi-use rooms, energy demand for heating and cooling is decreased. Storage spaces for the support of activities performed in the multi-use room increases the effectiveness of the room, and contributes to decreased energy consumption of the building. These spaces do not have to be maintained at comfortable temperatures, as long as damage to their contents is not incurred. Depending on details of construction and the season, storage spaces will maintain temperatures as much as 11° C (20° F) above or below room temperatures and are another factor in minimizing energy consumption.

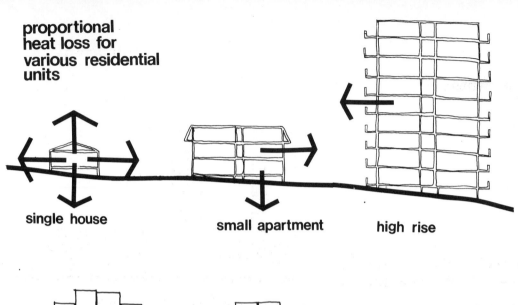

proportional heat loss for various residential units

single house

small apartment

high rise

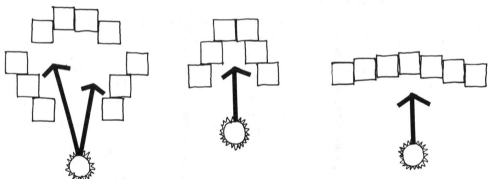

clusters and townhouses can be arranged to minimize heat loss and maximize winter solar collection

128

Common spaces are areas suited to recreational, social, leisure, and creative activities which require more space than is available in the individual living areas. People from a number of living areas share and support the common spaces. With this system the quality and diversity of the common spaces is greater than any individual could provide for himself. Of course, certain rules for use, conduct, maintenance, and repairs need to be established to the mutual agreement of all.

The cluster concept is easily adapted to the economical application of solar collectors for space heating, domestic water heating, or both. The type of collectors used would be a matter of economics and the needs the collectors are designed to satisfy.

For even further energy conservation, a number of living units can be consolidated under one building envelope, thereby reducing the wall area exposed to the outside air. Heat transmitted through interior walls can be used in adjoining living units and not lost to the atmosphere. The temperature difference across these walls will be so small that heat transmission will be kept to a minimum, stabilizing interior temperatures and decreasing energy consumption.

The average person in the US now uses more energy than the average citizen of any other country in the world. Up to 60% of the energy used in this country is wasted, partly due to poor planning of buildings and transportation systems. As the supplies of fossil fuels, which we depend upon for heating, transportation, and electricity diminish, increased prices will be paid. In some cases shortages of energy will be encountered. To ease the impact of these high prices and shortages, the American lifestyle will have to change. Greater respect for the natural environment will have to be realized along with a decrease in overall energy usage. Energy conserving space planning should take place at all levels of design, from small homes to entire cities. Appreciation for the urban core instead of the suburban sprawl must become a reality.

People generate varying amounts of heat when engaged in different activities. Because of this, various rooms can be heated to different temperatures. Also, various tasks can be performed at different temperatures. For example, sleeping areas can be kept colder than living rooms or dining rooms. Bathrooms can be heated in the morning and evening and left relatively cool the rest of the day. Certain rooms may not be used for days at a time and should not be heated when unoccupied.

At times the heat given off by lights and people may exceed the heat loss of a building. When this occurs no additional heating will be necessary, and excess heat can be used to warm other portions of the building. Any excess heat generated in one section should be captured and used as a supplementary heat source for other areas. Heat given off by washers, dryers, and other mechanical equipment can be stored for later use. The moisture from dryers can also be used to add humidity, and thus perceived temperature, to a space. In some large high rise structures, the south side during the winter may be air conditioned while the north side is being heated. Buildings should have controlled zones for heating and cooling, so that the excess heat from one area can be used to warm another. This would decrease the wasteful operation of systems.

The shape of a building is a factor to consider when planning the placement of insulation. The envelope of a multistory building will have a higher percentage of wall area and a lower percentage of roof area than a sprawling single or double story shopping center. Additional roof insulation in the multistory building would not control as large a percentage of the total heat flow as it would in the roof of the shopping center.

In multistory structures wall insulation is a primary consideration. It should be continuous through the height of the building, including between stories. Too often insulation is provided only between the ceiling and floor of a given story and not between a ceiling and the floor above. These areas may comprise 50% of the wall area and should be insulated.

Overhangs, shading devices, and recessed windows in tall buildings decrease the internal heat gains, yet allow for year-round natural illumination. Without these, the direct solar gains can be excessive, and energy consuming mechanical equipment has to be provided to handle the heat. It may be practical in some cases to use heat pumps to distribute excess heat to areas throughout the building.

On many commercial buildings a system of heat pumps, water tanks for heat storage, and cooling towers would be practical. Using heat pumps serving various zones, heat can be extracted from areas with southern exposure, transferred to storage, then distributed to areas requiring heat. The cooling towers would be used for evaporative cooling of the storage tank water during the hottest summer months.

The location of rooms within a building should be determined with consideration being given to the amount of heat which activities within the room will generate and the amount of heat which will be lost because of the room's location. Spaces which are not occupied a great deal of the time, such as corridors, closets, mechanical equipment rooms, laundries, and garages, can be kept at lower temperatures. They can be located on the north side of a building to provide buffer zones between cold north winds and occupied areas.

Rooms should be oriented to microclimatic conditions. During the day the north and east walls receive the least amount of radiation; they therefore remain the coolest. South walls receive radiation throughout the day, while west walls are exposed to intense late afternoon summer sun. Walls with these last two orientations will be the hottest which may significantly contribute to heat gains.

Areas in which heat is generated, such as laundry rooms or mechanical rooms, should be located on the north or east sides of a building so that heat can either be vented to the outside in the summer or used to warm the inside in the winter. By locating vents on these cooler sides, the hot air is most easily vented. Before hot air is exhausted, using it to preheat domestic water or to supplement other heat sources should be considered.

The north side of a building is a good location for a kitchen, hall, stairway, closet, or other areas that are not continually occupied. Internal heat is generated in kitchens by stoves, ovens, and other appliances, which warms the kitchen in the winter and may be used to initiate cooling inductive ventilation in the summer.

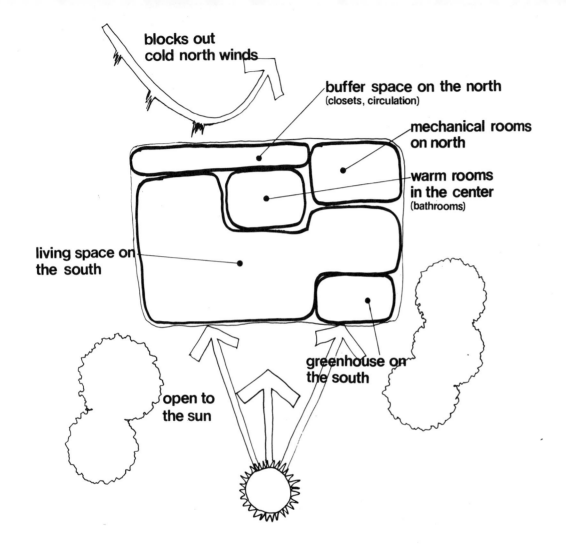

blocks out cold north winds

buffer space on the north
(closets, circulation)

mechanical rooms on north

warm rooms in the center
(bathrooms)

living space on the south

open to the sun

greenhouse on the south

The south side of a building is a good location for living, family, and dining room type areas. These may be adjacent to greenhouses or contain some type of beneficial as well as decorative foliage. Heat gains and illumination to these areas are greater than to any others, making them ideal for continuous occupancy. Entries on the south are functionally compatible with these rooms.

Bathrooms may be placed at the interior of an arrangement of rooms, and the walls of the bathroom thermally insulated from adjacent spaces. Once heated, a closed bathroom will remain warm until the door to the outside is opened and outside air allowed to infiltrate. If equipped with its own clock thermostat, a bathroom can be warmer than surrounding rooms in the morning and evening, and left relatively cool through the rest of the day.

The position of interior rooms with respect to exterior spaces is important to optimizing energy conservation. Patios, gardens, trees, bushes, and ground cover play an important role in regulating exterior temperatures adjacent to buildings. They can add humidity to the air while lowering temperatures through evaporative cooling and shading. This conditioned air can be drawn into a house by inductive ventilation and used to cool indoor temperatures. Patios provide pleasant outdoor areas for relaxation, dining, and social activities, and have their temperatures influenced by nearby buildings as well. A massive wall will retain daytime heat and reradiate that heat at night which can be a benefit, especially in the spring and fall, to nearby outdoor areas.

Space use and planning are extremely important in commercial structures. The effectiveness of leased spaces can be greatly improved by providing room layouts in which storage spaces to support activities have been carefully planned, and in which rooms are arranged within the building according to function. This is true of low rise, medium rise, and high rise structures; each has unique characteristics which must be considered as a consequence of their individual designs, shapes, volumes, dimensional ratios, and types of occupancy.

Support spaces for offices, reception areas, conference rooms, and retail spaces will increase the utility of each of these. As a consequence, tenants will need to lease a smaller total area, decreasing their costs, and lowering vacancy rates. Energy conservation will result from maximum utilization of space, high population density, and the sharing of resources (heat) in properly zoned structures.

Lifestyles, attitudes, and design philosophies will have to be altered if total energy consumption is going to be reduced. A simple procedure like turning off a light bulb in an unoccupied area is a start, but more drastic changes are necessary; for example, permanent improvements in the efficiency of the light bulb itself. Many of these will come about as more people come to the realization that supplies and recoverable reserves of fossil fuels are limited. Changes in lifestyle do not necessarily mean a decrease in comfort. Suggestions presented here are intended to increase the comfort of all.

groupings of retail or office buildings
listed in order of increased thermal efficiency

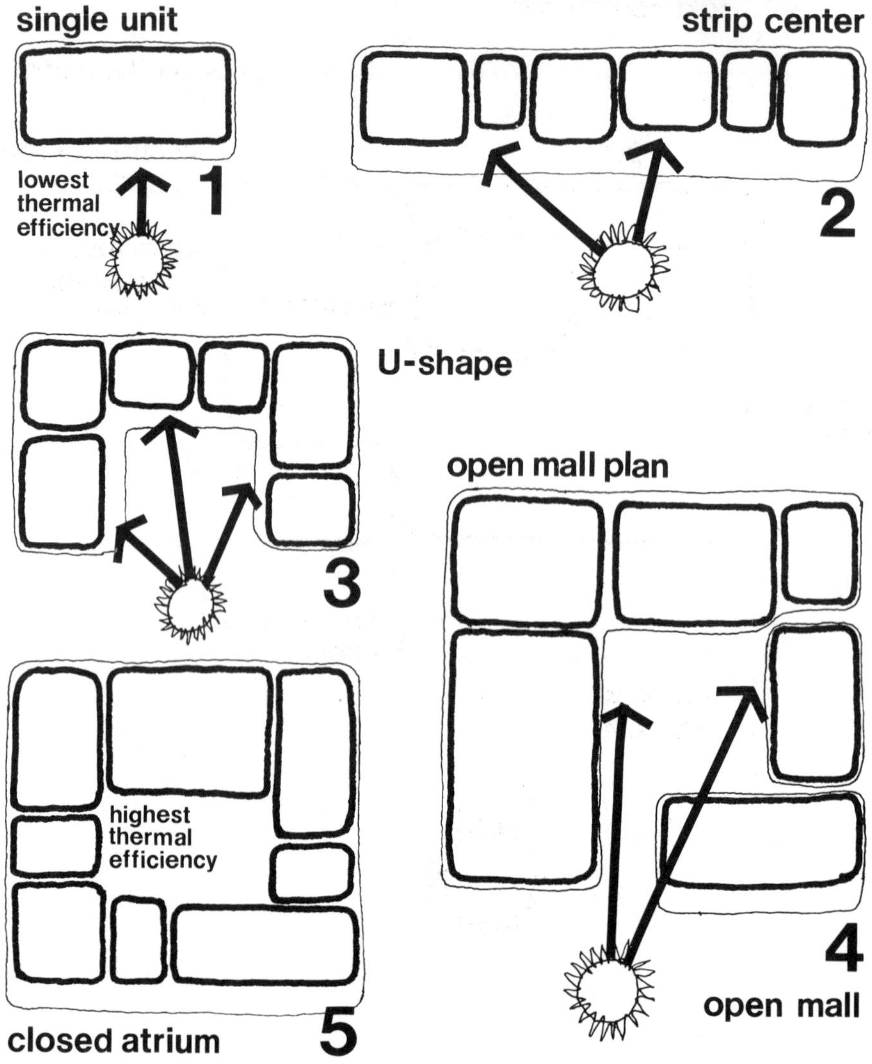

single unit

lowest thermal efficiency

1

strip center

2

U-shape

3

open mall plan

open mall

4

highest thermal efficiency

closed atrium

5

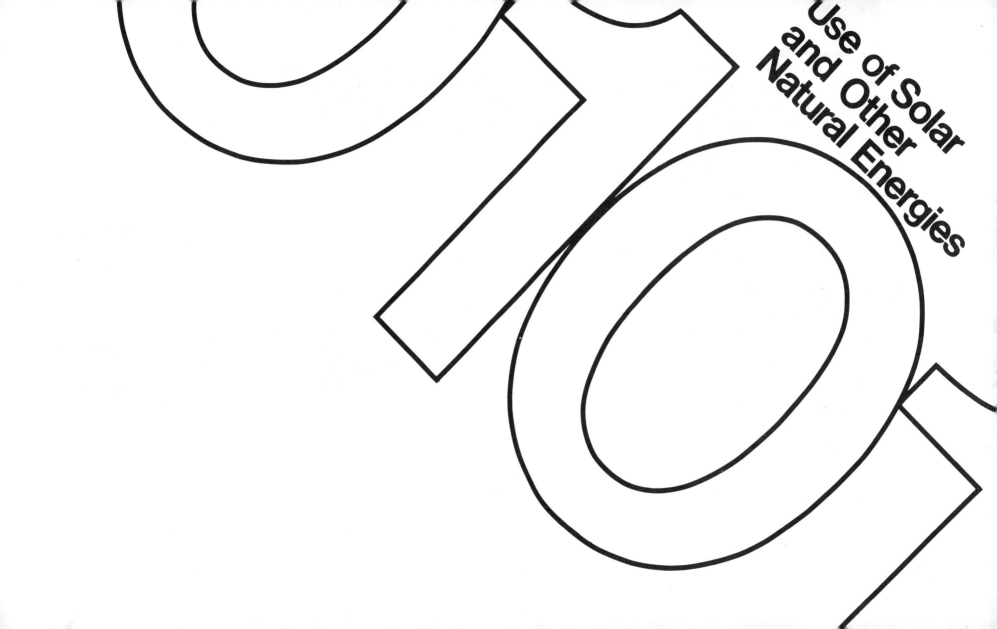

Use of Solar
and Other
Natural Energies

In order for the manmade environment to use the natural energy available to it most efficiently, it must be planned with consideration given to materials selection, site and building orientation, microclimatic conditions, landscaping, and many other factors. The environmentally responsive building should have minimal negative impact on its site, maximize efficient use of all energy, maximize human comfort and effort, and promote human talents by providing a pleasing atmosphere in which to live or work.

To economically incorporate alternative energy devices into buildings, energy conserving measures must be taken which diminish the total energy usage. Energy conservation in some buildings can reduce energy demand by as much as 80%. Many different items relating to energy conservation should be considered and evaluated for possible use in all buildings. Areas in which energy conserving practices can be employed have been divided into eight categories and are presented in this chapter as follows:

a. Site and Climate
b. Entrances
c. Walls and Roofs
d. Windows and Other Openings
e. Lighting
f. Ventilation
g. Mechanical Systems
h. Appliances

Each aspect of a building should be planned for its best utilization of all energy. Many of the design ideas presented in this chapter include passive systems which use the natural energies available from the sun, wind, water, and earth.

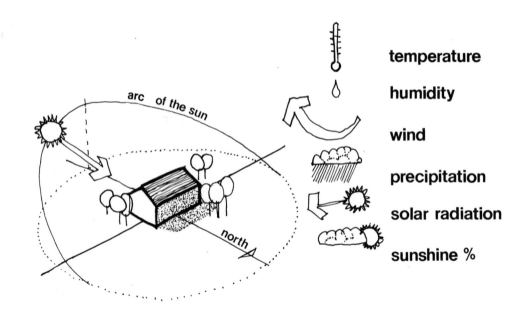

temperature

humidity

wind

precipitation

solar radiation

sunshine %

site and climate

To properly consider the placement and orientation of any building on a site, its regional and micro-climatic conditions must be evaluated. The micro-climate is peculiar to each individual site and results from the influences of surrounding buildings and landscaping, topography, soil structure, and ground cover, as well as regional geographic influences.

Local climatic conditions must be evaluated in conjunction with regional weather data to determine the microclimate of a particular site. The regional data are usually available from the ASHRAE Handbook and the U.S. Department of Commerce, which maintains weather monitoring stations throughout the country. Data that should be considered which influence the design of a building are:

1. The daily and monthly average dry bulb and wet bulb temperatures.

2. The relative humidity (the percent of water vapor in the air relative to the maximum amount it can hold).

3. The wind speed and direction; daily, monthly, and annually. (A wind rose map will provide the velocity of the wind and percentage of wind flowing in each compass direction.)

4. The precipitation (the quantity, type, and duration of all snow, rain, hail, etc.).

5. The incident solar radiation (usually measured in Langleys, where one Langley is defined as one calorie per square centimeter (3.69 Btu/sq. ft.), on a horizontal surface.

6. The seasonal percentages of sunshine (the percentage and probability of sunshine will help to predict monthly heating loads and storage capacities needed).

Plans for using alternative energy sources require that foresight be used in site selection and building placement. If an urban site directly north of a high rise building is selected, it is unlikely that much direct solar radiation will be available at the site for collection.

In hilly or mountainous areas, a site with southern exposure is most advantageous for year-round control of indoor climatic conditions. These will have a favorable orientation for solar collection in winter, and the hill itself will offer partial protection from heat-robbing winter winds. Sites on the north side of hills should be avoided, except in climates where heat gains during all seasons need to be controlled.

The proper orientation of a building on its site is dependent upon the regional climate and the micro-climate. To optimize energy use, it would be desirable for a building to have the greatest amount of area with southern exposure during the winter (to maximize solar heat gains when they are most necessary), but the least amount of area with this same exposure during the summer (to minimize solar heat gains when unnecessary). It is obviously difficult for the proportions of a building to change with the seasons, so a compromise is needed to balance the annual heat gains against the annual heat losses, in determining the optimum building proportions based on thermal considerations only.

The chart at the right (adapted from "Design with Climate" by Victor Olgyay) shows various ratios of north-south dimension versus east-west dimension for different types of climates, with the optimum and an acceptable range of building proportions for each.

Changes in orientation can be effectively implemented by modifications in siting and design. Trees and shrubs can be used to reduce solar heat gains in the summer. Deciduous trees (trees which shed their leaves at the end of each growing season) provide shade in summer months and allow sun to pass through in the winter. Various trees provide differing degrees of shade, depending on leaf structure and density. Many trees allow diffuse light to penetrate, permitting natural lighting levels to be maintained, while others are practically opaque.

Deciduous trees, though effective for providing shade, are not so valuable as windbreaks. Therefore, on the north or northwest side of a building (depending on the prevailing cold winter winds) coniferous trees should be used. Their use and type depend on the density of the branch structure, how close to the ground they grow, and their height at maturity. Their effectiveness as windbreaks is governed by the proximity of individual trees. It is necessary to direct the air flow over the trees instead of around them.

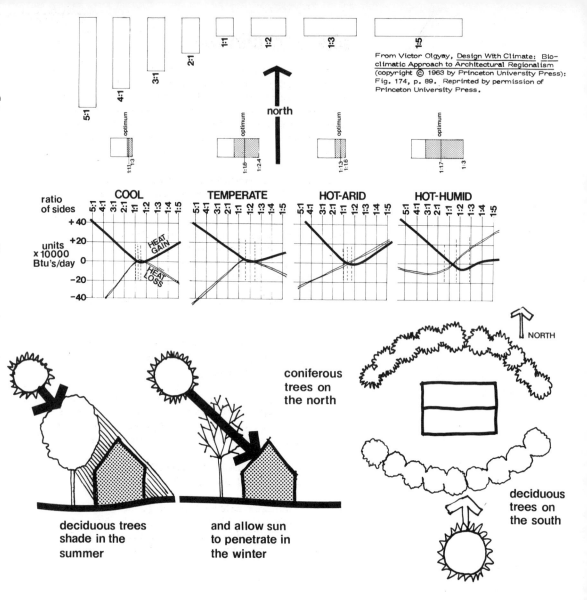

From Victor Olgyay, Design With Climate: Bio-climatic Approach to Architectural Regionalism (copyright © 1963 by Princeton University Press): Fig. 174, p. 89. Reprinted by permission of Princeton University Press.

north

ratio of sides

units x 10000 Btu's/day

COOL   TEMPERATE   HOT-ARID   HOT-HUMID

HEAT GAIN
HEAT LOSS

deciduous trees shade in the summer

and allow sun to penetrate in the winter

coniferous trees on the north

deciduous trees on the south

NORTH

136

From Architectural Graphic Standards by Ramsey and Sleeper, Copyright © 1970 by John Wiley & Sons, Inc. Reprinted by permission of John Wiley & Sons, Inc.

inside forest

outside forest

inside forest			outside forest		
ground temperature (F)	air temperature (F)	humidity	ground temperature (F)	air temperature (F)	humidity
60°	75-80°	77%	70°	90°	85%
45°	65-70°	75%	50°	75°	70
33°	58-60°	60	30°	60°	60
	44-48°			45°	
	29-31°			30°	
	17°			15°	

The dead air space behind trees can act as insulation space. With coniferous trees on the north and west and deciduous trees on the south and east sides of a building, maximum protection from the winter cold winds, minimum obstruction of the sun's warmth in the winter, and maximum shading from the sun in the summer can be realized.

The use of trees can also assist in the insulation of a building against both heat gain and heat loss. To the left is a comparison of temperatures and humidity inside and outside a wooded area. It is evident that trees produce conditions which moderate the extremes in both hot and cold temperatures.

Plant areas also help to purify the air. As air moves through the leaves of a plant, carbon dioxide in the air is exchanged for oxygen from the leaves. The plant structure also causes air turbulence, permitting particles suspended in the air to fall to the ground.

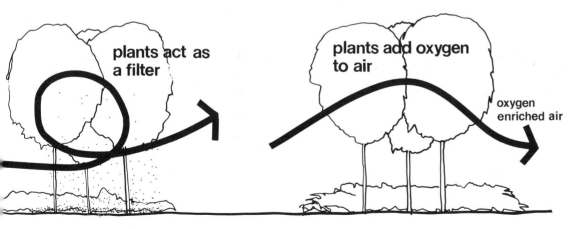

plants act as a filter

plants add oxygen to air

oxygen enriched air

carbon dioxide

oxygen

Earth can be used to minimize the amount of exposed surface area of a building. Mounds of earth (berms) on the north side can considerably reduce the heat loss in that area. Prevailing winter winds (which usually come from the north or northwest) will carry away heat faster from an exposed north wall than

from any other exposed building surface. Therefore, it is best to minimize the exposed wall surface
area on the west and north sides.

Earth is effective as an insulator below frostline. A
mixture of mulch and soil can decrease the depth of
the frostline because it is an insulator. An insulative
ground cover, such as bark or leaves, also aids in
diminishing cold penetration in the soil.

Berms can be useful in directing noise and snow
away from a structure. Sound cannot penetrate the
mass of a berm and is either absorbed or reflected
by it. By tilting the berm's surface as shown, the
sound is reflected upward away from the building.
Proper positioning and forming of berms will direct
winds, causing snowdrifts to form away from buildings and entrances.

One method for preventing excessive exposure of a
building to the elements is to place it underground.
In the opinion of Malcolm B. Wells, New Jersey
architect, underground architecture offers a more
ecological alternative approach to building design.
He has, in fact, completed a number of underground
structures, one of which also uses solar heating.
These buildings restore and intensify the natural

 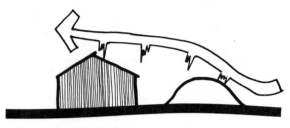

cold north winds directed over the buildings by berms

**noise reflected upward
by berms**

**snow drifts can be diverted
by the use of berms**

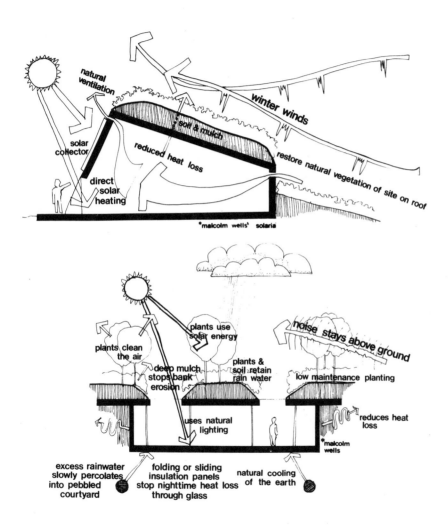

natural
ventilation

winter winds

soil & mulch

solar
collector

reduced heat loss

restore natural vegetation of site on roof

direct
solar
heating

*malcolm wells' solaria

plants use
solar energy

noise stays above ground

plants clean
the air

deep mulch
stops bank
erosion

plants &
soil retain
rain water

low maintenance planting

uses natural
lighting

reduces heat
loss

*malcolm
wells

excess rainwater
slowly percolates
into pebbled
courtyard

folding or sliding
insulation panels
stop nighttime heat loss
through glass

natural cooling
of the earth

landscaping of their area. Mr. Wells uses the chart shown below, entitled "MAN'S WORKS compared to the miracle of WILDERNESS in fifteen ways....", to evaluate buildings. A score from −100 to +100 can be attained in each of fifteen categories. If the total score is negative, the building is "dead"; if it is positive, the building is "alive".

	-100	-75	-50	-25	0	+25	+50	+75	+100	
destroys pure air										creates pure air
destroys pure water										creates pure water
wastes rain water										stores rain water
produces no food										produces its own food
destroys rich soil										creates rich soil
wastes solar energy										uses solar energy
wastes fossil fuels										stores solar energy
destroys silence										creates silence
dumps its wastes unused										consumes its own water
needs repair and cleaning										maintains itself
disregards nature's pace										matches nature's pace
destroys wildlife habitat										provides wildlife habitat
destroys human habitat										provides human habitat
intensifies climate and weather										moderates climate and weather
destroys beauty										beautiful

139

Fences and other external structures can be used in a variety of ways to deflect winter winds over or around structures, decreasing the heat loss of a building.

In the summer months, heat loss resulting from winds is not important, but heat gain is. Ventilation and circulation of the air within a building are of primary concern. During the summer, low velocity winds generally come from the south and southeast. These air currents can be used for encouraging summer ventilation. For maximum efficiency of air flow and heat gain/loss, a building in the northern latitudes should block out cold winter winds from the north and west and encourage the infiltration of cooling summer breezes from the south.

Earth structures built to the north side of a building can be used to reflect radiation into northern windows and against northern walls. White rock covering the earth will provide diffuse reflected radiation, while with a mirrored surface a predictable reflection is possible. These site modifications direct heat and light into alcoves which are in the building's shadow, and would otherwise receive no direct heating or illumination.

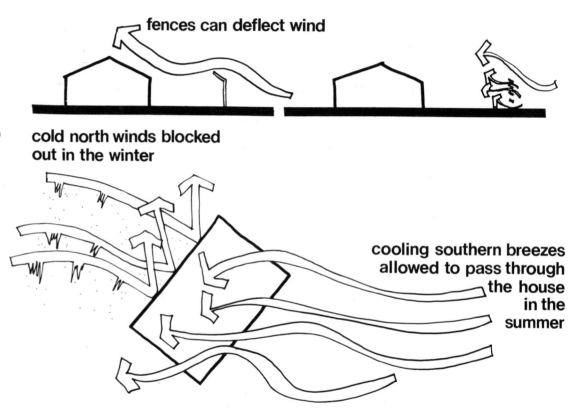

fences can deflect wind

cold north winds blocked out in the winter

cooling southern breezes allowed to pass through the house in the summer

Some suggestions for landscaping and site modifications to improve comfort and conserve energy are listed below.

1. Plant deciduous trees and vines on the south side of the building to decrease summer heat gains, yet allow the winter sun to enter. Conifers planted on the north side will act as an effective windbreak, decreasing winter heat loss.

2. Fences of the proper design will provide privacy, security, sound control, and wind control.

3. Where few or no openings are present on the north side of a house, the earth can be banked against the side to provide insulation. The height of the embankment will depend on the site, configuration of the house, plus structural and waterproofing details.

4. Berms strategically placed on a site will direct wind flow and control sound. These and embankments should be surfaced with ground cover.

5. Shrubs can effectively be used to produce cool ventilation intake air, as a consequence of the shade they cast near ground level.

6. Patios, offering outdoor areas to comfortably spend leisure time, may also be used to temper indoor temperatures. Patios with white gravel on the south can be used in cold climates to reflect heat to interior spaces or occupants. Deep trenches around patios filled with gravel and moistened with water will reduce temperatures through evaporation.

7. Manure and natural compost from yard clippings are good fertilizers. Most commercial fertilizers have a petroleum base, hence should be used only if compost and manure are not available.

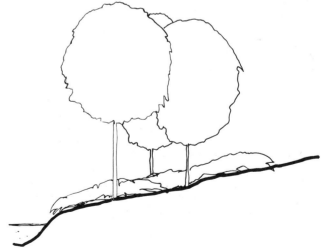

entrances

Controlling heat loss and gain through entrances is a major concern. All edges around entrances leak air and correspondingly lose or gain heat. To minimize these heat transfers doors must seal as tightly as possible at the sides and at the bottom threshold. The edges must set square with a weatherstrip to prevent air movement. Weatherstripping can usually provide adequate protection in this area. Winds will carry away heat at a greater rate than still air outside a door. To keep air as still as possible, wingwalls and landscaping can provide windbreaks. One of the major and immutable problems with an entrance is that each time it is opened a great quantity of cold or warm air enters the adjoining room. It is possible for the entire volume of air in a room to be exchanged when the adjoining door is opened. This problem can be intensified if the entrance is oriented to intercept the prevailing winter winds.

To prevent large heat losses due to infiltration when a door is opened, airlocks can be provided. An airlock has two doors and an air space between; one opens to the exterior and one to the interior. When the exterior door is opened, only the air in the airlock can escape; therefore, only a small quantity of heat escapes. The exterior door is then closed before the interior door is opened, allowing only a small amount of the interior heat to escape to the airlock. This prevents the entire room from being infiltrated with hot or cold air. Generally, airlocks are not heated, but naturally maintain a temperature between that of the exterior and the interior temperatures.

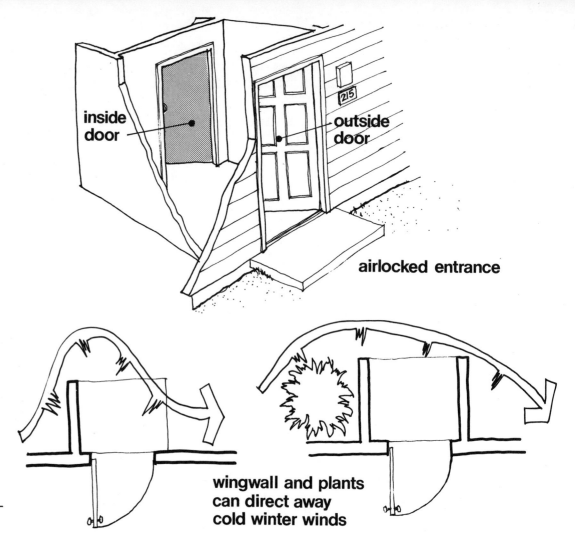

inside door

outside door

airlocked entrance

wingwall and plants can direct away cold winter winds

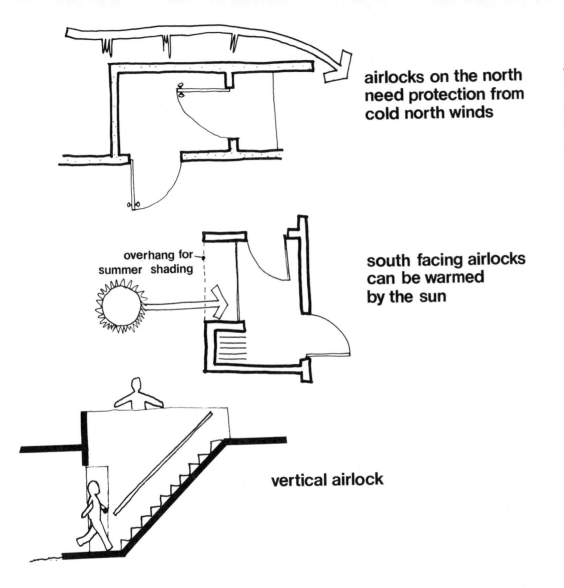

**airlocks on the north
need protection from
cold north winds**

overhang for
summer shading

**south facing airlocks
can be warmed
by the sun**

vertical airlock

Airlocks work most effectively when placed on the
south or east sides of a building. Wingwalls must be
included with airlocks which are placed on the west
and north sides to prevent prevailing winds from
decreasing their effectiveness. Airlocks on the
south can be warmed by the sun. With insulated glass
on the south side and a floor constructed of a heat
retaining material, the airlock unit can stay warm
several hours after dark.

Airlocks can be used for more than just passageways.
They can be used for storage of unheated items
(sleds, snow shovels, umbrellas, etc.) or as areas
to remove debris (mud, snow, etc.) before entering
the building.

Another entrance which is effective for reducing heat
loss is the vertical airlock. Since cold air will not
rise, and warm air will not fall, an entrance located
below a building will allow the least possible amount
of heat to escape. When the door is opened a small
amount of warm air will exit and be replaced with
cold air. The infiltrated air will not rise since it is
colder (and heavier) than the air in the upper room.
As the person who entered ascends into the heated
space, he is greeted by the layer of warm air.

A revolving door works as an airlock because each
chamber traps and releases only a small amount of
heated air to the outside each time it completes a
revolution. The edge, top, and bottom seals are of
major importance to keep this type of door effective.
One problem which is encountered is that as the
tightness of the seal around each chamber is improved, the force needed to rotate the door and
overcome the friction increases. This places a

limit on how effective the door can be as a heat loss barrier, but such doors are often successfully used in large building entries as windbreaks.

In snowy regions, provision should be made to prevent snowdrifts from obstructing entrances. Reflective surfaces can be used above or around the door area to accomplish this. These should be calculated to allow entryways to be heated by the sun during the winter months and to be shielded from the sun during the summer. Care should also be taken to avoid directing the solar radiation into the eyes of people when they are exiting or entering. The reflection should be directed toward a flooring surface dark in color, to absorb as much energy as possible. Also, adequate drainage must be provided so that melted snow doesn't freeze in the entryway at night.

A garage can act as a climatic buffer when so located to attenuate north and northwest cold winter winds, or extended to protect an east or west entry. Radio controlled garage openers conserve energy in cold climates by decreasing the time that the door needs to be open for passage of the vehicle.

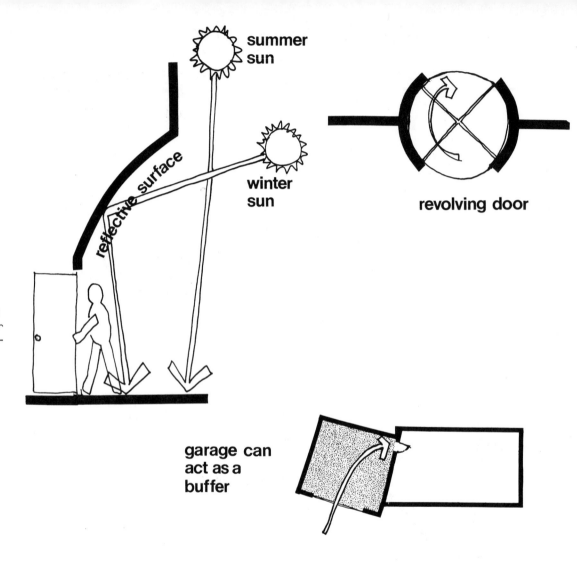

summer sun

reflective surface

winter sun

revolving door

garage can act as a buffer

weatherstripping reduces infiltration

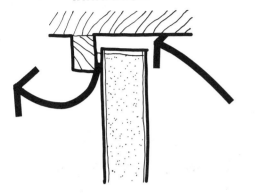

insulated doors lose less heat than non-insulated doors

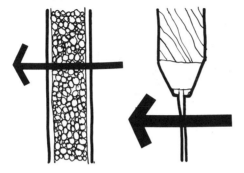

solid insulated door

outside louvered door for security and ventilation

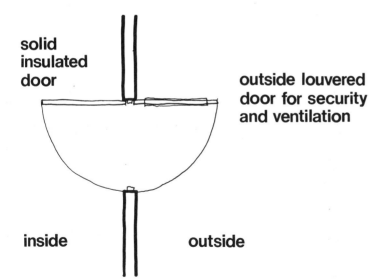

inside   outside

Some hints for improving the energy efficiency of entrances are presented below.

1. Weatherstrip around all doors to decrease infiltration. Continuity is important, especially at corners.

2. Storm doors can serve as additional insulation panels. Be certain these fit tightly enough to create a dead air space.

3. Avoid using entrance doors against which the cold winter wind directly blows.

4. A porch can be added to a house to serve as an airlock entry.

5. Wingwalls can be added to shelter entrances from prevailing winds.

6. A dark colored south facing door will absorb winter heat, but may be too hot in the summer unless sheltered by overhangs from higher sun altitudes.

7. During cold or hot weather avoid having doors open for extended lengths of time.

8. Use insulated exterior doors.

9. Avoid the use of mail slots that lose energy and allow infiltration.

10. A weatherstripped, screened, and louvered door with a lock for security may be used for ventilation and should be designed so that an insulated panel can be added in winter.

145

walls and roofs

Walls and roofs form barriers which separate the inside of a building from the outside environment. These barriers resist the passage of most elements, but are not impervious to the flow of air and moisture, with their accompanying heat. Methods for calculating the rate of heat transfer were presented in Chapter 8. Ideally, walls and roofs are designed and constructed to minimize winter heat loss and summer heat gain, while maximizing winter heat gain and summer heat loss from interior spaces. Two important properties which regulate heat flow are the thermal resistance (insulation) and thermal mass (thermal inertia) of walls and roofs.

To minimize the rate at which heat is lost or gained through exterior surfaces, they should be constructed of materials having high thermal resistance. Generally, the higher the thermal resistance, the slower the heat flow through the wall or ceiling, with a corresponding decrease in interior heat loss or gain. One component contributing to the thermal insulation value of an element is the layer of air on its inside and outside surfaces. Surface texture has an influence on the thickness and stability of these layers of air. Shake shingle exteriors produce a thicker, more stable layer of air than metal or other smooth surfaced materials.

Any material which is in contact with the inside and outside surfaces of a wall or roof aids in the transmittal of heat through it. Wall studs or floor and ceiling joists can be staggered to produce discontinuities in wall construction which act as thermal barriers. This system can also be used to thicken wall cavities to produce greater space for insulation without adding larger studs.

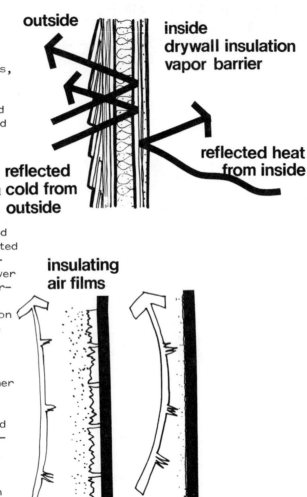

outside | inside
drywall insulation
vapor barrier

reflected cold from outside

reflected heat from inside

insulating air films

insulation reduces heat loss

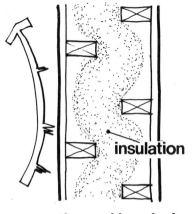

thermal breaks in the unit

brick acts as a heat sink

outside | inside

Most building materials are porous and allow moisture-laden air to penetrate through them. Since latent heat is required to evaporate the moisture into the air, a moisture loss also represents a heat loss. When the moisture condenses, it may cause discoloration, rotting, or peeling of exterior paint. A vapor barrier around all wall and ceiling areas will significantly decrease air penetration and heat loss. The barrier should be positioned on the "warm" side of the wall so that the adverse effects of condensation are minimized. In cold climates the vapor barrier should be applied near the inside surface, preferably between the drywall and studs. In hot humid climates it should be applied under the exterior skin material.

heat builds up in the wall during the day

heat is reradiated at night

By virtue of their mass, all materials have the ability to store heat. The amount of heat that can be stored per unit mass varies from one material to the next. The thermal mass of building materials is a very important property which must be considered in the total design process.

Certain materials such as brick, block, stone, and adobe can store large amounts of heat. In their traditional position on the exterior of a building, these materials in summer will absorb heat for long periods of time before allowing it to flow to the interior. This produces the "time lag" effect and completely alters the heat flow which is generally calculated using the U-value alone. Later in the day when the outdoor air temperature is lower than the internal wall temperature, the heat stored in the walls will be conducted to the outside air. The thickness of the wall, density of the material, and its heat capacity govern these effects.

Brick veneer can be used most effectively to moderate interior temperatures if it is applied to the inside rather than the outside of a building. In the winter months, the brick absorbs heat during the day or at any time when the temperature of the air in the room is greater than the temperature of the brick. When the air temperature drops so that the temperature of the brick is higher, the brick will give heat back to the room by radiation and convection, helping to maintain a stable temperature. Insulation with a reflective barrier should then be placed on the exterior of the wall where it is most effective. Summer operation of this system is similar to the process described above but produces an even greater time lag.

A house design by the French engineer Felix Trombé integrates the thermal storage capacity of concrete into a flat plate collector. Components include a thick concrete wall facing south and painted black, plus a sheet of glass covering the wall to decrease convective heat losses and to form an air chamber between the wall and the glass. Vents at the top and bottom of the wall connect the air chamber to the adjacent room. As the sun warms the wall, the air next to it heats up and rises, passes through the top vent into the room, and allows cold room air to enter the bottom vent by displacement. The cold room air is warmed as it passes next to the wall and exits the collector at the top. The concrete wall also absorbs heat which is conducted directly into the room.

By opening a vent between the glass and the outside, the collector works as a ventilator in summer. Air warmed by the sun flows up the collector and out the vent. Cool air enters the room by displacement through vents located on the north side of the building.

Roofs can also be designed to retain or reflect heat energy. Harold Hay has successfully tested a system which uses sealed clear plastic bags filled with water, backed with black plastic, and located on a roof to store heat. The bags are supported by a structural corrugated metal roof through which heat can be transmitted. During winter days the sun heats the bags. Most of the heat is stored in the water and a portion is conducted through the roof and radiated to the room below. During winter nights the bags are covered with a movable insulated panel and the heat stored during the day radiates downward into the space. During a summer day the insulation reflects the heat of the sun, and at night any heat stored in the water is radiated upward to the sky.

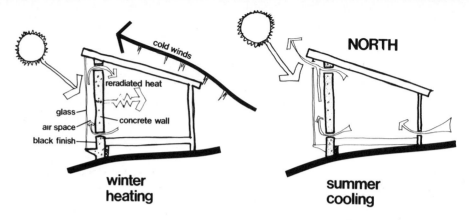

winter heating

summer cooling

Trombé house

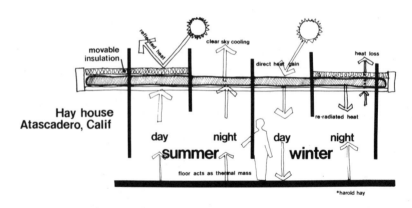

Hay house Atascadero, Calif

day night day night
summer winter

The presence of water on a roof can help moderate solar heat gains into a building by acting as an evaporative coolant. Simple ponding on the roof can also cool its surface and may extract heat from the interior. In the winter, however, a roof pond may lose more heat by evaporation in 24 hours than it gains during 8 hours of daylight.

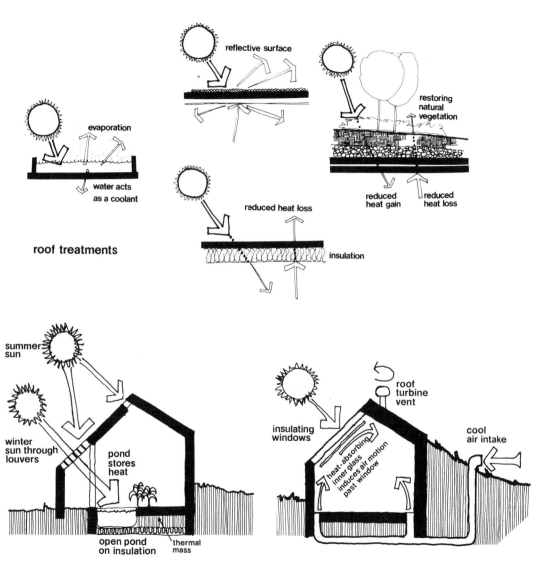

evaporation

water acts
as a coolant

reflective surface

reduced heat loss

insulation

restoring
natural
vegetation

reduced
heat gain

reduced
heat loss

roof treatments

summer
sun

winter
sun through
louvers

pond
stores
heat

open pond
on insulation

thermal
mass

roof
turbine
vent

insulating
windows

heat-absorbing
inner glass
induces air motion
past window

cool
air intake

Similarly, earth or sod on a roof help to moderate heat loss and gain while sustaining natural vegetation. This minimizes the negative impact of the building on the environment.

A mass of water may be incorporated into a building design as an open pool, swimming pool, or closed tank to act as either a stabilizer of internal temperatures or as a reservoir of thermal energy. With just a few degrees rise in temperature a large water mass can hold substantial amounts of useful energy. Solar, wind, and earth energies all work well in combination with water.

A relative humidity of 50% is regarded as physiologically ideal for most human beings but may not be high enough for some plants. Open bodies of indoor water, in supplementing ambient humidity, can cause problems by evaporating excessively at high room temperatures. This can be turned to advantage by lowered room temperature, in which human comfort is more easily sustained.

Earth temperatures in most geographic locations remain fairly cool and constant in temperature. During the warm and hot seasons of the year, air can be naturally circulated through underground areas to reduce its temperature, then brought into a building to sustain indoor comfort. This action can be maintained by employing the sun to increase the upward convection of air through roof stacks.

Interior thermal masses of earth, concrete, masonry, or their combination with a well insulated building envelope make the building itself a storage system of thermal energy. Moisture should be kept away from insulation, as water content will reduce or nullify its efficiency.

149

The properties of walls and roofs are affected by their surface colors, with dark tones absorbing more heat than do light tones. Mirrored or white surfaces can reflect a great deal of radiant heat from any source, including the sun. Combinations of reflective and absorptive materials can be used to direct solar radiation, and to provide diffuse lighting.

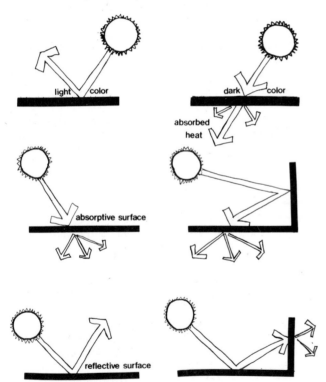

There are a number of modifications that can be made to walls and roofs which improve energy conservation. Some of these are presented below.

1. Increase the insulation in the ceiling. A heat resistance value of 30 (R-30) should be a minimum standard for the insulation in the ceiling. (Note: This and other resistance values are federal recommendations. Cost effective use of solar energy as a primary space heating source may suggest, in cooler climates, far higher thermal resistances for roof, floor, and walls.

2. Increase the insulation in floors above basements and crawl spaces. A heat resistance value of 19 (R-19) should be a minimum standard for insulation in the floor.

3. If possible, increase the insulation in outside walls to a resistance value of 13 (R-13). Insulation in the upper portion of walls is most important.

4. If possible, increase the mass of outer walls. An application of masonry veneer would increase the thermal inertia and time lag of the wall.

5. Close attic and crawl space vents in the winter if doing so won't interfere with combustion air to the furnace.

6. When remodeling, place vapor barriers between the drywall and studs. This will decrease infiltration.

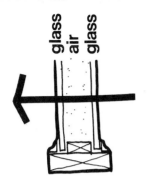

double glazed windows

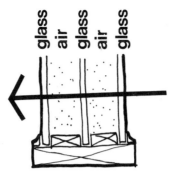

triple glazed windows

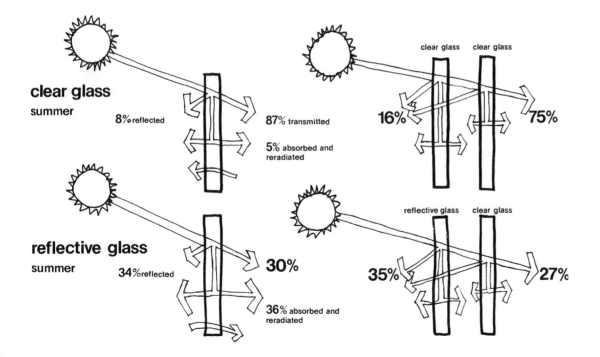

clear glass

summer

8% reflected

87% transmitted

5% absorbed and reradiated

reflective glass

summer

34% reflected

30%

36% absorbed and reradiated

clear glass   clear glass

16%   75%

reflective glass   clear glass

35%   27%

windows

A great deal of heat is lost from a building through windows and other openings. Beneficial heat from the sun can also be gained through windows. To minimize energy loss through glazed areas, insulating glass which decreases heat conduction can be used. Insulating glass consists of two or more panes with a dead air space between. In the best prebuilt units, two panes are edge-welded after the air separating them is partially evacuated. Within limits, the insulating quality of the air increases as the distance between the panes increases. Optimum distance is 3/4"; beyond that the velocity of convection currents formed between the panes of glass will transport large amounts of heat from one surface to the other. In some extremely cold climates, three panes of glass can be used which produce two insulating air spaces.

The location of windows on a building will have an influence on their size, shape, and type. On the west and south sides of a building it may be advantageous to use glass with a reflective surface to shield out the intense summer sun. To the left is an illustration showing the approximate percentage of heat transmitted and absorbed by different glass types and configurations.

A true dead air space is a good insulator but loses its effectiveness when convection currents propagate in it. One interesting system which uses a material other than air was patented by Steve Baer of Zomeworks and is called "Beadwall" (Trademark). This system has two panes of glass spaced several inches apart. During the day the space or chamber between the glass contains air only. At night, expanded

151

polystyrene pellets are blown into the chamber, displacing the air and forming a heat barrier.

Sliding, magnetically clipped, or hinged insulated panels can be used to decrease heat flow through glass. During winter, these would be closed at night, decreasing heat loss, and opened during the day to allow sun penetration (except north) to interior spaces.

There are exceptions according to latitude, but window area should generally be kept to a minimum on all sides of a building except the south, which should offer the maximum possible consistent with energy conservation. The winter sun should be used to heat a building during the day and the windows should be sufficiently insulated to prevent heat loss at night. Window recesses or overhangs can be designed to allow the sun to enter a window in the winter months and shade the window from the intense summer sun.

While determining the size and placement of windows, many things should be considered. One of the more important variables is the position of the sun, which is continually changing, not only throughout each day but seasonally as well. Two angles are therefore needed to specify the position of the sun in the sky: one, the azimuth angle, which is a measurement in degrees of the sun's position in plan view (looking down from above); the other is the altitude, which is a measurement in degrees of the sun's elevation above the horizon. At a given date and time, both the altitude and azimuth differ according to earth latitude.

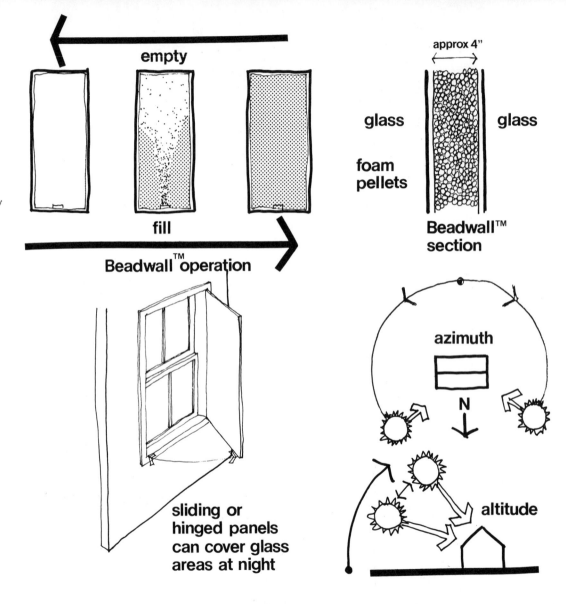

empty

fill

Beadwall™ operation

glass   glass

foam pellets

approx 4"

Beadwall™ section

sliding or hinged panels can cover glass areas at night

azimuth

N

altitude

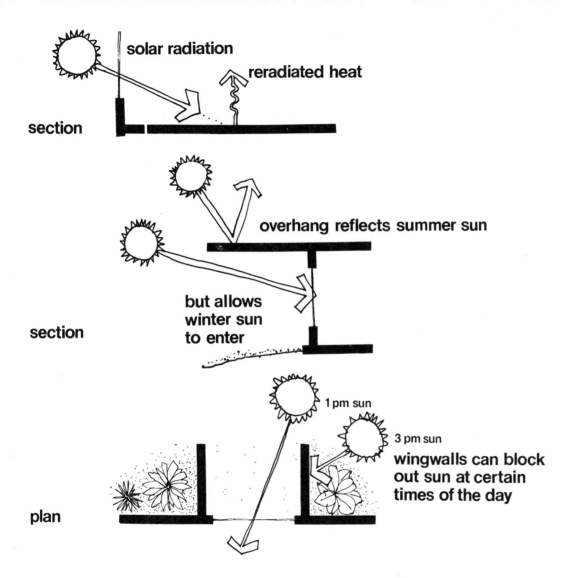

section

solar radiation

reradiated heat

section

overhang reflects summer sun

but allows
winter sun
to enter

plan

1 pm sun

3 pm sun

wingwalls can block
out sun at certain
times of the day

Direct solar radiation can be intercepted and stored by the interior floors and walls of a room, then reradiated to the room. Windows which extend to the floor line allow the floor to be heated by direct radiation. The heat will be absorbed and stored in floors of massive construction (concrete, masonry, rock, etc.) and released when room temperatures start to fall.

Interior walls can be used to reflect diffuse light into a space. A long, narrow window is greatly enlarged in its effect by being located adjacent to a wall or ceiling of light color, while softly diffusing the glare component of direct sunlight. Using the sun for natural diffuse lighting can also replace significant amounts of energy required for artificial lighting. In all buildings, natural lighting in some areas could be adequate for direct occupant use.

Overhangs and wingwalls can serve to reflect or block the sun's rays; the reflected light can be used to provide natural lighting to interior spaces not receiving direct solar radiation. Proper design and placement of wingwalls and overhangs can provide rooms with warming winter sun while controlling the entry of intense summer sun.

To optimize the design of buildings to be energy conserving, windows should:

1. have overhangs or wingwalls to block out the summer sun

2. allow the winter sun to penetrate and warm interior spaces

153

3. allow the heating of floors and walls which can retain and reradiate heat

4. be recessed, to trap heat and reduce convective losses

5. prevent heat loss at night and during days of limited sunshine

6. provide natural lighting for interior spaces

Because gaps exist around movable sections even when they are closed, more heat is lost through operable windows (ones which open) than through fixed windows. To allow the least possible heat loss, the use of operable windows should be kept to a minimum, and air outside a window should be regulated for natural air circulation induced by the sun and wind. Insulated windows provide dead air spaces which decrease the amount of heat reaching the outer glass surface.

Large glazed areas can be protected by buffer spaces. A greenhouse area on the south will act not only as an insulating buffer, but as a source of heat for the building. As the sun heats the greenhouse, excess heat energy can be vented into the house during the winter and to the outside in the summer. At night the greenhouse acts as insulating air space for the southern windows. During especially cold periods, the greenhouse will be tempered by the heat loss from the house, tending to maintain safer temperatures for plants.

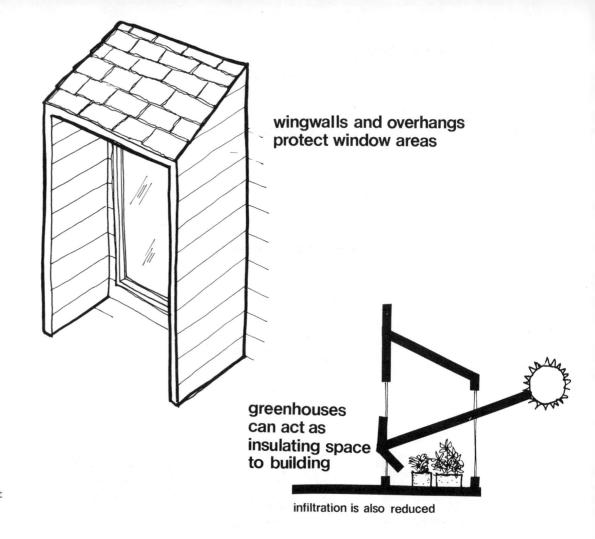

wingwalls and overhangs protect window areas

greenhouses can act as insulating space to building

infiltration is also reduced

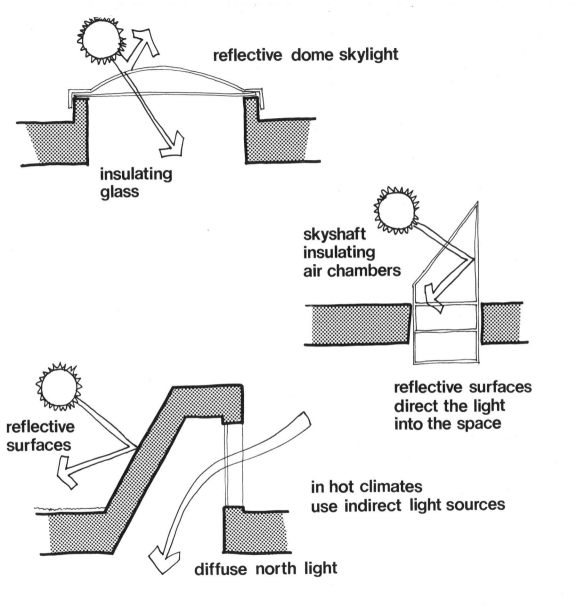

reflective dome skylight

insulating
glass

skyshaft
insulating
air chambers

reflective surfaces
direct the light
into the space

reflective
surfaces

in hot climates
use indirect light sources

diffuse north light

Another device which will allow light to enter a space, yet decrease internal direct solar gains during all seasons that might overheat the interior, is the sky-light. Normally, a great deal of heat is lost through skylights, because the hottest air will be at the high-est point in a room, which is generally the location of a skylight. Recent developments in double and triple layers of glass and plastic for skylights with insulating air spaces between the layers have de-creased the heat loss problems. The benefit of sky-lights is diminished if they are placed on a shaded north side of a building, which should usually be avoided.

"Skyshafts", as developed by Richard L. Crowther, are composed of several sealed air chambers which are lined with white reflective sides and clear lenses at the top or horizontal positions. These provide an interior space with diffuse lighting.

In southern latitudes, conventional skylights should be used sparingly, as summer heat gains through them can be disastrous. A north-facing monitor window, as shown below, would be a more protected light source in such regions.

Skylights used in conjunction with sloping mirrored surfaces will illuminate and heat areas not generally receiving direct radiation. This is especially impor-tant in providing light and warmth to rooms with northern exposure.

Another feature which will provide natural illumination to rooms not adjacent to the exterior of a building is an atrium – an open shaft or hole extending vertically the full height of the building, forming an interior courtyard at the bottom. The walls of the atrium may be surfaced with relatively large window areas to allow as much light as possible to enter adjoining spaces which, without the atrium, would be closed to daylight. Sliding doors of insulated glass can provide access to the courtyard. The effect that wind will have on the air in the atrium is controlled by the size, shape, and proportions of the atrium, the edge configuration between the roof and atrium walls, and the wind velocity. Ideally, wind will blow over the atrium, allowing the air against the windows to remain still, affecting a minimization of heat losses. This effect will also carry snow over the atrium and prevent its accumulation on the atrium floor. At times, vortices or other undesirable wind conditions may form in the opening to the atrium which disturb the air inside. These effects, however, are not easily predicted.

windows can be protected against cold winds by an atrium

Some suggestions for improving window areas for energy conservation are presented below.

1. In the winter make sure all windows are closed tightly, and caulk or weatherstrip all edges not operable.

2. Fasten storm windows of glass or plastic to exterior frame to increase thermal resistance. Metal frames must also be insulated, as they can conduct or convect large amounts of heat.

3. Keep storm windows in place during the summer on all windows except those to be used for ventilating.

4. Install draperies or shades on all windows and use them as movable insulating barriers to control year-round heat flow. These should fit tightly to sill or floor and be made of good thermal materials.

5. Open windows on warm but not hot days to allow natural ventilation, before resorting to powered air-conditioning.

6. Attach external shading devices such as overhangs, sidewalls, recesses, awnings, or lowered sun screens to reduce solar heat gain.

7. Coat west facing windows with reflective films, and/or install mirror chromed venetian blinds inside.

8. Replace metal window frames with wooden ones.

lighting

Proper lighting of buildings is of major importance, accounting for up to 50% of the total energy expended in some nonresidential buildings. Lights produce heat which in a large building can be a substantial amount, either useful for heating, or an additional load against cooling.

Lighting inevitably affects the biophysical responses of man. Certain wavelengths of light help the eye to maintain proper levels of the chemicals necessary for vision. Lighting can also have the psychological effect of altering a person's mood, attitude, or efficiency.

Artificial light is generally produced by converting electrical energy to light energy, using a lamp. Some light is produced by the direct burning of fuels, as exemplified by a candle or kerosene lamp. For the most part, the light produced in this way is used to create an atmosphere rather than for illumination.

A measure of the intensity of light is the number of "lumens" it produces, and the efficiency of a light source is expressed in terms of the number of lumens of light it emits per watt of electrical power consumed. Energy conservation in lighting can be accomplished in a number of ways; the simplest is to change from one light source to another. Different light sources have different electrical conversion efficiencies, light output, lamp life, and color temperature, expressed in degrees Kelvin. Below are some typical efficiencies for different types of lamps.

Lamp Type	Output
Incandescent	19 lumens/watt
Mercury Vapor	57 lumens/watt
Fluorescent	75 lumens/watt
High Pressure Sodium (yellow)	125 lumens/watt
Low Pressure Sodium (orange)	183 lumens/watt

By using the most efficient lamp which will provide the necessary light intensity and color, electricity consumption can be kept to a minimum. As the power rating of a lamp decreases, so does its efficiency; hence a 15 watt lamp will produce fewer lumens per watt than one rated at 100 watts. Operating a lamp at a lower voltage than specified will decrease its light output but will disproportionately prolong its life. The light output can also be affected by the cleanliness of the bulb or tube, as well as the type of maintenance program for a building and its light fixtures.

Light output, along with fixture design and surface reflectivity of walls and ceilings, affects the amount of useful light striking a surface.

Another important consideration in the selection of a lamp type is the color it emits. Lamps do not equally produce all the colors of the spectrum; their spectral content varies greatly from one type to another. Incandescent bulbs give off a warm reddish

157

light. The three types of fluorescent bulbs; cool white, warm white, and daylight have unique spectral qualities. The color content of cool white is predominantly yellow and green, while warm white is red-orange. The color emitted by daylight tubes varies from one manufacturer to the next, but all attempt to simulate natural outdoor light. Some manufacturers have, in fact, developed fluorescent lamps showing 91% chromaticity, against a natural daylight index of 100%. By contrast, average cool white fluorescent lamps yield an index of 68%.

The life of a lamp is expressed as the number of hours it will provide useful light (figured as a statistical average). This can be an important consideration when specifying a type of lamp. On large projects the replacement of burned out lamps can be a major long term expense in terms of labor and the cost of replacement bulbs. Fluorescent and sodium lamps offer up to 36,000 hours life, or between 10 and 50 times the longevity of incandescent lamps.

Other factors which influence the quality of light at the work surface are the design of the light fixture (luminaire) containing the lamp and the reflectivity of walls and ceiling. Fixtures should be designed to deliver the maximum possible useful light to a work surface. Their placement should be such that glare and shadows are minimized. Reflective walls and ceilings can help to lower the lighting requirements of a space. Dark colored walls, ceilings, and floors will absorb light which could be reflected and used for illumination of a task area.

usually light bulb packages include the useful light in hours and the average light output in lumens

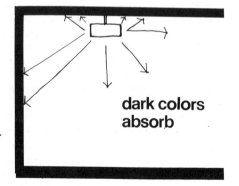

dark colors absorb

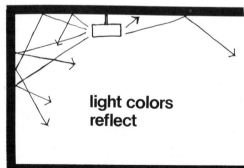

light colors reflect

Existing light fixtures can be relamped to great advantage and with no wiring alterations, in the following categories:

Intermittent or task lighting
Incandescent (tungsten filament) lamps should contain krypton gas, which can increase lamp life by a factor of five, at lower wattage per lumen output.

General indoor lighting
Incandescent (tungsten filament) lamps should contain krypton gas, which can increase lamp life by a factor of five, at lower wattage per lumen output. Fluorescent lamps should also contain krypton gas, which lengthens lamp life while maintaining the same lumen output at a 9% lower wattage level. High pressure sodium, even more economical, may be used where color rendering is not critical.

Outdoor area lighting
For small areas, self-ballasted mercury vapor, high pressure sodium, or krypton/tungsten incandescents can be used; all of these offer longevity and lumen output superiority over incandescents. Large areas are best lighted by low-pressure, low-wattage sodium (LPS), which offers outstanding lumen/watt output and lamp life. Numerous large American cities have recently made partial LPS street lighting or freeway installations, recognizable as a warm yellow-orange glow. This color rendering produces monochromatic night vision and high object resolution; that is, all objects which under mercury vapor lighting appear black or indigo blue are instead highly visible as shades of gray. The improvement is immediately noticeable in cities using red fire trucks, for example, as these no longer appear black at night.

Task lighting
Lighting levels should meet, not exceed, the requirements of the tasks or activities to be performed in a space. In the past, general overall lighting of high brightness was utilized to provide equal intensity levels over an entire room, but much of the light was not productively utilized. The amount and quality of light required to perform a given task varies from one activity to the next. The concept of task lighting involves providing an area with an amount and quality of illumination sufficient to perform the task for which that area is intended. Quality of light is extremely important, and in many cases it is a more imperative criterion than the foot candle level (quantity). Task lighting makes more sense than equal lighting throughout a building. Lighting fixtures over individual desks use much less energy than is used illuminating an entire office area. Since light intensity varies as an inverse function of distance from the source, the closer a fixture is located to the task area, the less light it must emit to provide the necessary illumination.

Photocells can be used to switch on artificial lighting when natural lighting is inadequate or to prevent outdoor lighting from coming on until it is dark. They can also be used to activate shading devices when sunlight is too intense.

provide task lighting instead of general illumination

In some buildings all of the lights in a large room have to be activated in order to turn on the lighting fixture over one desk. If the light fixtures could be turned on independently, a great deal of energy could be saved. It is not always practical for each light to be independent. Energy could be saved if zones, composed of a small number of fixtures, could be turned on at once.

Listed below are measures which can improve the efficiency of the lighting in an existing building, contribute to energy conservation and a reduced electricity bill, yet maintain minimum lighting standards necessary for seeing and safety.

1. Turn off all lights when they are not needed. One 100 watt incandescent bulb burning for ten hours requires 11,600 Btu's of energy to be liberated at the generating plant.

2. Use fluorescent lamps in suitable areas as they produce more lumens per watt and have a longer life than incandescent bulbs.

3. Keep lamps and lighting fixtures clean. If they are dirty, light output can be significantly reduced.

4. Light colors used for walls, rugs, draperies, and upholstery will reflect more light than dark colors and reduce the wattage of artificial lighting fixtures required.

5. Consider installing solid state dimmer switches when replacing light switches. They reduce energy consumption by permitting lamps to be operated at reduced power levels.

6. For any application it is better to use one large bulb instead of two equivalent small ones. As an example, one 150 watt bulb delivers 2,880 lumens while two 75 watt bulbs deliver only 2,380 lumens.

7. A three-way bulb should be at the lowest setting for watching TV. The higher settings can be used for tasks such as reading, writing, sewing, etc.

8. Do not turn off any fluorescent lamp which will be needed twice or more during a fifteen minute period; the shortened lamp life caused by switching will more than offset electricity saved in this manner. Fluorescent lamps are "ignited" by bursts of extremely high voltage, which affects lamp and ballast (transformer) life according to the number of "starts". Incandescent lamps, by contrast, deteriorate according to total operation time, with the number of starts being less important. These lamps should be turned off whenever unused.

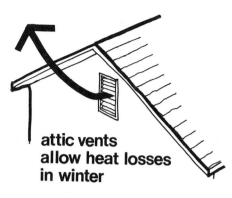

**attic vents
allow heat losses
in winter**

ventilation

As air moves or circulates it gains heat from objects warmer than itself and loses heat to colder ones. Wherever two air masses of differing temperature mix, however, the larger will prevail in its effects. During cold weather it is therefore important to prevent air from carrying heat out of a building, while during the summer maximum ventilation is encouraged, unless mechanical air conditioning is used.

Tight seals on all the windows and doors of a building minimize the infiltration of cold air and the loss of warm air. Weatherstripped doors and inoperable (fixed) windows also prevent excessive heat losses due to infiltration. Cracks in the construction around windows, doors, and other openings can leak air, particularly in masonry buildings. (See Chapter 8 for methods for calculating heat loss due to infiltration.) Caulking and sealants can be used to stop infiltration heat losses around joints and openings, while vapor barriers prevent air and moisture penetration through walls, floors, and ceilings. Any exhaust fan, bathroom vent, or other opening for air circulation should be provided with a damper to minimize heat loss when not in operation.

Vents which are used year-round, such as direct ducted bathroom fans and kitchen exhaust fans, can remove undesirably large amounts of heat from a space. Ductless filtered fans are available which remove odors from room air, then return it into the room directly. Air is pulled into one such unit by a fan, then filtered through a natural citrus-based chemical that absorbs odors caused by bacteria and mildew. The air and its heat are then returned to the room. Exhaust fans can also be equipped with

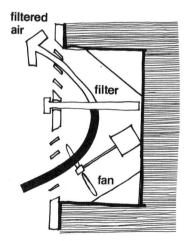

filtered air

filter

fan

**ductless fan does not
exhaust heat from rooms**

heat exchangers which remove heat from the air before it is exhausted to the exterior.

Any fans or vents which are not equipped with dampers allow heated air to escape almost continuously. All equipment of this type should be fitted with dampers which are designed to seal as tightly as possible to minimize heat loss. Dampers should be insulated and spring-loaded toward their interior (room) side, or otherwise controllable. In some applications they may be remotely located and require motorized control.

During the winter, heated air will leak out of a building through its attic vents. Doors or shutters can be attached to attic vents which are closed in the winter and opened in the summer. Cracks around hatch doors or stairways leading to an attic are pathways through which heated air travels on its way out of the building and should be sealed with weatherstripping or caulking. Access hatches themselves should be fully weatherstripped, with insulation equal to that of the surrounding attic.

Ventilation should be minimized during cold months and encouraged during warm months. However, minimum ventilation standards should always be met to provide adequate air supply. (A little known but well established fact is that with all windows and doors closed, a typically constructed building is provided all necessary air changes purely by infiltration.)

The air required for ventilation varies depending on the activities performed in a space and whether or not smoking is allowed. If smoking is permitted, up to nine times more air must be brought through a space hourly to maintain necessary purity levels than if smoking were prohibited. This can mean oversized mechanical equipment to move and purify the air. A more reasonable alternative might be confinement of smoking to certain small areas of any building. Interior building materials will deteriorate at a greater rate in an environment where smoking is allowed. The air pollutants from smoking can even cause disorders in sensitive electronic equipment such as computers and activate smoke detection alarms.

Natural ventilation of a building will initiate passive cooling during the summer. The exhaust of excess warm air and the intake of cool air help to lower interior temperatures. To facilitate the exit of warm air, dampered vents should be located at high points in a building. If natural thermal pressure differentials do not produce sufficient flow velocities, fans, turbines, or plenums can be used to accelerate them. Attic fans will, if needed, more quickly remove hot air that accumulates during the day. Operable vents, connecting the building space with the attic, provide a pathway for the upward flow of hot air. The motion of outside air can often be used to induce interior air movement without the aid of fans. Wind can be used to power a turbine or directed in such a way that pressure changes result which move inside air.

Buildings themselves act as large scale ducts and should be designed with options for unobstructed movement of air. Openings should be sized to permit natural ventilation to occur. Air contained in

162

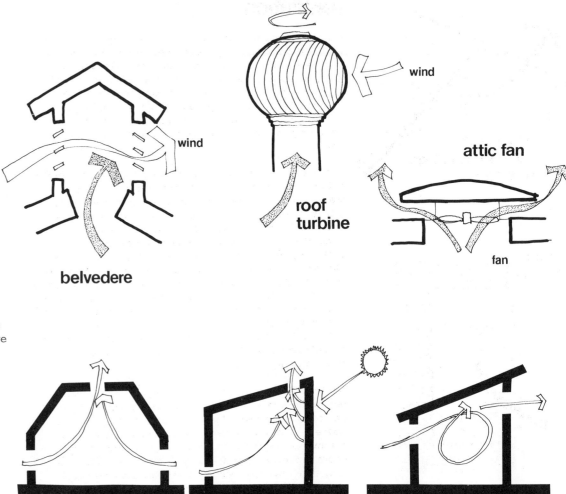

belvedere

wind

roof turbine

attic fan

fan

buildings act as ducts

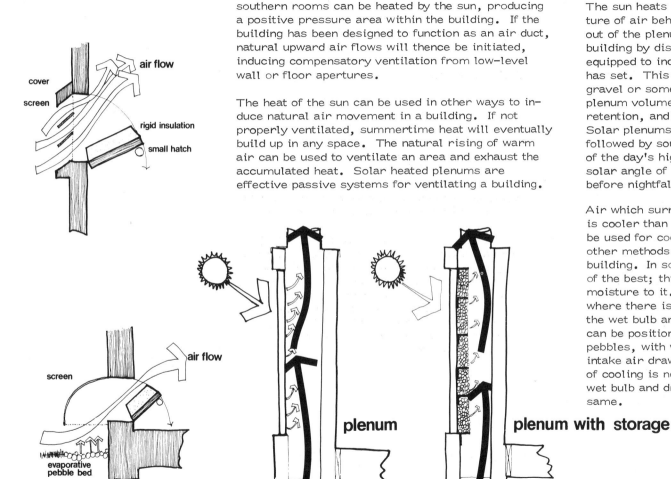

southern rooms can be heated by the sun, producing a positive pressure area within the building. If the building has been designed to function as an air duct, natural upward air flows will thence be initiated, inducing compensatory ventilation from low-level wall or floor apertures.

The heat of the sun can be used in other ways to induce natural air movement in a building. If not properly ventilated, summertime heat will eventually build up in any space. The natural rising of warm air can be used to ventilate an area and exhaust the accumulated heat. Solar heated plenums are effective passive systems for ventilating a building.

The sun heats the plenum and raises the temperature of air behind a cover. Warm air rises up and out of the plenum, allowing cooler air to enter the building by displacement. The plenum can be equipped to induce air movement even after the sun has set. This is accomplished by incorporating gravel or some other heat storage material into the plenum volume, for daytime heat absorption and retention, and dispersal of heat the following night. Solar plenums are most effective on west exposures, followed by south. In these locations they make use of the day's highest ambient temperature, maximum solar angle of incidence, and latest radiation period before nightfall.

Air which surrounds trees, shrubs, or ground cover is cooler than air away from vegetation, and should be used for cooling summer ventilation. Several other methods can be used to cool the air entering a building. In some locales, evaporative cooling is one of the best; this method cools the air by adding moisture to it. Evaporative cooling works well where there is an appreciable difference between the wet bulb and dry bulb temperatures. Intake vents can be positioned near ground level over a bed of pebbles, with water running over the pebbles and the intake air drawn off immediately above. This form of cooling is not effective in humid climates, as the wet bulb and dry bulb temperatures are nearly the same.

Some suggestions to improve the energy efficiency of existing buildings, relative to ventilation, are listed below.

1. Insulating batts are available with foil vapor barrier backing. These can be used to decrease heat losses and infiltration.

2. Install recirculating fans with charcoal or citrus base filters in kitchens and bathrooms, as they decrease the amount of heated air piped directly out of the building in winter.

3. Use bath and kitchen ventilating fans which exhaust to the outside only when needed.

4. In the winter, be sure that fireplace dampers are closed except when the fire is going.

5. Modify the fireplace to deliver more heat to the building. Heat exchange methods are noted in Chapter 8.

6. On warm days, open windows for ventilation instead of using air conditioning or fans.

7. Install controllable vents between the various levels of a building. Basement temperatures can be drawn up for cooling. Vents can be closed and heat retained at the level where activities are being performed.

8. Install a turbine vent at the highest point of any pitched attic roof to draw hot air out of the attic during summer. The same vent should be tightly closed during winter. Eave venting or other vents are required to provide incoming displacement air.

9. In the summer, use the fireplace flue as a vertical stack to initiate passive ventilation.

10. Provide vents and a stack to allow natural ventilation of hot air in greenhouses.

11. Heat transfer wheels, heat pipes, and other heat exchangers can conserve energy by transferring exhaust air heat to incoming cooler exterior air.

12. Exterior outdoor air is often polluted. Closed cycle air quality reconstitution through filtration or by use of plants conserves the need to temper varying temperatures and eliminate the pollution of outdoor air. Indoor smoking usually renders the indoor air quality below that of polluted outdoor air. Health, efficiency, and quality of the environment is threatened by smoking and air pollution.

mechanical

Mechanical systems use energy to heat, cool, and ventilate the interior environment of a building. In the heating mode, fuel or electricity is used to warm the air or fluid which circulates in a mechanical system. Exhaust gases from furnaces or boilers can carry up to 50% of the system's heat energy up and out the flue. This need not be wasted; most can be extracted from the exhaust gases and be used elsewhere in the building. To accomplish this, heat exchangers have to be added to the exhaust ducts; they may contain a fluid or air which is pumped through a series of tubes. As hot exhaust gases rise through the flue, they flow around the tubes, eventually giving up their heat to the cooler fluid inside. The exhaust gases still exit from the flue, but only after giving up some of their heat in this manner. A series of heat exchangers could be used in a fireplace flue to effectively capture the energy given off during the burning process.

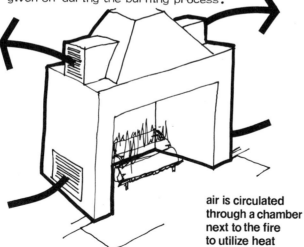

air is circulated through a chamber next to the fire to utilize heat

The successful operation of many of the systems which have been presented depends upon transferring heat from one fluid to another or from one location to another. Heat pumps can perform this function, but they require an input of work and rely upon moving parts to accomplish their function. Two other types of devices which transfer heat are heat exchangers and heat pipes.

Some heat exchangers have moving parts while others do not. A heat wheel is one that slowly rotates to transfer heat and must be powered by a motor. In the winter it extracts heat from exhaust air, to temper incoming cold air. In the summer, heat from incoming warm air is transferred to existing cool air precooling the entering air.

The wheel, packed with aluminum or stainless steel wool, rotates between an exit duct and an entry duct, picking up heat on the warm side and rejecting it on the cold side. These have been designed to transfer not only sensible heat, but latent heat of evaporation. Moisture is absorbed from the humid airstream by a desiccant, lithium chloride, and is released to the dry airstream.

The most common heat exchangers are ones which have no moving parts and transfer sensible heat only. These are exemplified by automobile radiators. A hot fluid, air or liquid, is piped into the heat exchanger where it flows into a network of small tubes, greatly increasing its surface area. A cooler fluid circulates on the opposite side of the tubes, and heat is transferred to it by conduction. The rate of heat flow depends on the relative temperatures of the two

165

fluids, their heat capacities, flow velocities, surface areas, plus the thickness and type of material out of which the heat exchanger is constructed. Transfers of heat from air to-liquid or from air to air are generally less efficient than transfers from liquid to air or from liquid to liquid.

A device which is considerably more efficient for air-to-air heat transfers is the heat pipe. This is a simple object composed of a sealed tube, containing a working fluid and a wick. When one end of the tube is heated, the working fluid evaporates, absorbing large amounts of heat, and moves to the cooler end of the tube. There it condenses, giving up the heat, and flows back through the wick to start the cycle over. A pipe one inch in diameter and two feet long can transfer 12,500 Btu an hour at 982 C (1800 F) with only a 10° C (18° F) temperature drop from one end to the other.

The heat pipe was invented in the 1940's by Richard S. Glauger, a General Motors engineer, who used it in refrigerators. It has been used in many of the most recent spacecraft but has had only limited use in homes and buildings. Principal among the reasons for its not being used more in homes and buildings is its high cost. Hopefully, the cost will decrease when these devices are manufactured in quantity.

Heat exchangers can be used in several different ways to prevent "waste" heat from leaving an interior space. Fluid exchangers can remove heat from water in shower and tub drains, dishwashers, water or refrigerator condensors. This heat can be used to preheat domestic water or to supply heat for

a space heat pump. Heat exchangers are available in a variety of sizes and configurations, but their purpose remains the same, to transfer heat from one medium to another.

A heat pump is a device which is able to produce more energy as heat than is contained in the fuel with which it is operated. This may seem to be a contradiction of the laws of thermodynamics, but is not, for the energy input comes from two sources, the operational fuel (mechanical energy) and the natural (ambient) energy of the environment. Its name derives from the ability to "pump" converted ambient energy into a space to be heated, or pump "heat" out of a space to be cooled.

The operation of a heat pump takes advantage of the principles of latent heat. A working fluid, refrigerant, circulates throughout the heat pump system which gives off heat when it condenses and picks up heat when it evaporates. The heat content of the refrigerant is affected more by changes of state than by changes of temperature.

In a normal cycle, refrigerant is compressed to a relatively high pressure, producing a temperature increase. Mechanical energy is required to operate the compressor to raise the internal energy absorbed by the refrigerant at low ambient temperature to a temperature level useful for space heating. For Freon, the most common refrigerant, this temperature will reach approximately 60 C (140 F). Hot gaseous refrigerant is pumped through a heat exchanger where it loses some of its thermal energy, cools and condenses to a liquid. This is called a condensation heating cycle; while changing from a gas to a liquid, latent heat is given off.

After it has condensed, the fluid is still at a high temperature and pressure. By suddenly reducing the pressure using an expansion valve, the temperature can be reduced below -17.8 C (0 F). The low pressure refrigerant is then a mixture of liquid and

gas and is piped to a second heat exchanger where it takes on heat energy and vaporizes. This second unit is called an evaporator. Energy can be taken from "cold" environments since the refrigerant temperature is below -17.8 C (0 F). At this point the cycle starts over with the gaseous refrigerant being condensed.

If the hot gas is pumped to a condenser located in a building, the system is operating in the heating mode. If it is pumped to a condenser located outside the building, it is operating in the cooling mode.

The advantages of this system are that it is reversible and that the heat energy output at the condenser is greater than the mechanical energy input to the compressor. The second energy input to the system comes when the low pressure, low temperature liquid refrigerant travels through the evaporator in which it vaporizes and takes on ambient heat from the outside. Heat pumps are energy conserving because the heat energy output is generally 2.5 to 5.0 times greater than the fuel input. As earlier noted, COP expresses this ratio.

Heat pumps that draw their heat of evaporation from air, and release heat through a condenser to air at another location are called air-to-air heat pumps. Heat can also be drawn from water and released to air, or vice versa, by air-to-water heat pumps. The ground can also be a heat source or sink and used in a ground-to-air or ground-to-water heat pump system. Each of these has advantages and disadvantages which should be considered for specific applications. Manufacturers of heat pumps should be contacted for this information.

air·to·air heat pump system

heat is removed from outside air and transferred inside

heat is removed from inside air and transferred outside

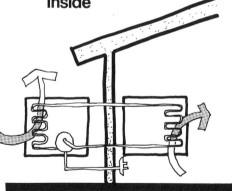

winter

summer

Because of the numerous adjustments and modifications that can be made to the mechanical systems of homes and buildings, these suggestions for energy conservation are divided into groups of listings for each of the different types of systems.

space heating

1. Lower thermostat setting to 20 C (68 F) during the day and 15.5 C (60 F) at night. A clock thermostat automatically lowers the setting at night, then raises it again in the morning. One of these may be a good investment.

2. Have the furnace and entire heating system serviced once a year, preferably in the fall to keep it operating at top efficiency.

3. Clean or replace the filter in a forced air heating system once a month. The fan will operate less if it doesn't have to force the air through a dirty filter.

4. Dust or vacuum radiator surfaces frequently. Dust and dirt will block the heat, reducing efficiency.

5. Consider installing a humidifier. Moist air helps you feel as comfortable at 21.1 C (70 F) with 35% humidity as you would at 23.3 C (74 F) with very low humidity.

6. Keep draperies, carpets, and upholstered furniture from obstructing heat outlets or radiators.

7. Don't heat to comfortable temperatures rooms which aren't being used. Close the doors leading to these areas and restrict the flow of heat through registers or radiators, or allow for cooler zones in large buildings.

8. During the day keep draperies and shades open in sunny windows; close them at night, as the radiative heat gain will nearly always exceed the convected heat lost to outdoor ambient air.

9. Move leisure furniture away from exterior walls to avoid cold drafts and the consequent lower perceived temperatures.

10. Keep fireplace dampers closed when the fire is not burning.

11. Provide combustion air from the outside directly to the fireplace and furnace. By not using heated indoor air for combustion, fuel bills can be significantly reduced.

12. Turn the thermostat down to 12.8 C (55 F) when the house or building will be unoccupied for a day or longer.

13. For comfort at lower indoor temperatures, the best insulation is warm clothing. Clothes made from natural fibers, wool and cotton, may be warmer than ones made from a similar thickness of synthetic materials.

14. Insulate all ducts which carry heated air and pipes which carry heated water.

space cooling

1. Set air conditioner thermostats no lower than 25.5 C (78 F). If humidity is controlled, this is a comfortable temperature.

2. Run air conditioners only on very hot days and set the fan speed on high. In very humid weather, set the fan speed on low to provide less cooling but more moisture removal.

3. In dry climates, consider installing an evaporative cooler. These devices generally use less energy than conventional air conditioners and cool an incoming air stream by extracting heat from it to evaporate water.

4. Have the air conditioning system serviced once a year, preferably in the spring, to keep it operating at top efficiency.

5. Clean or replace air conditioner filters once a month. Clean filters will decrease the amount of electricity used by the fan.

6. Do not air-condition unoccupied rooms. Close them off from the rest of the home or building.

7. Minimize the use of lights, appliances, sound equipment, and televisions. These all generate heat which increases cooling loads.

8. Shade windows with overhangs, wingwalls, or awnings, and deflect direct sunlight with light colored shading devices, preferably mounted on the outside of the building and having an air space for convection between them and the exterior wall.

9. Consider installing vents and exhaust fans to remove heat and moisture from attics, kitchens, and laundry areas and exhaust it directly to the outside.

10. Consider installing a wind-powered roof ventilator along the roof ridge and louvered grilles between the attic and ceiling, so that cool air entering basement vents will be induced by the exit of the hot attic air to flow upward through the building.

11. Locate outdoor cooling components behind opaque foliage where they will be shaded from the sun.

12. Insulate all ducts used to distribute cool air.

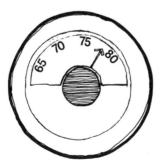

set air conditioning thermostat at 78° or higher

plumbing

1. Repair leaky faucets as quickly as possible. A hot water faucet leaking one drop a second wastes 700 gallons of hot water per year.

2. Keep the temperature control on the hot water heater set no higher than 60 C (140 F). Temperatures higher than this waste energy and shorten the life of a glass-lined tank.

3. Minimize water and energy consumption by running clothes washers and dishwashers only with full loads.

4. Make sure hot water storage tanks and hot water pipes are well insulated to decrease heat loss.

5. Take more showers than tub baths, because showers use less hot water.

6. Consider installing a flow restrictor in the showerhead which limits the flow to four gallons of water per minute, while allowing essentially unchanged water spray velocity.

7. Place a brick, or quart jar filled with water, in the tank of the water closet to decrease the amount of water used each time it is flushed. Don't decrease the volume too much or plumbing problems may result.

8. Do as much cleaning as possible, including the car, with cold water. This will save the energy used to heat water.

appliances

The analysis of homes and other buildings for optimized energy conservation requires that appliances be treated as part of the unified building system. Appliances are devices which perform work, provide heating or cooling, and require energy to accomplish their function. Because they require energy they generate heat which is transmitted by conduction, convection, and radiation to surrounding air. This heat can either contribute to space heating in the winter or contribute to space cooling loads in the summer.

The value of an appliance can be assessed in terms of a time-based energy audit of alternative methods for accomplishing the same end result. For example, a healthy human adult is able to open a can using a mechanical can opener in about the same length of time it takes to open a similar can using an electrical can opener. Much less energy is required to accomplish this task by hand than by the electrical appliance, and since the times involved are similar, the choice should be to not use the appliance.

Variable conditions may enter into the judgment concerning the value of an appliance. For example, another choice may be whether or not to dry clothes in an automatic dryer. If the weather is sunny and dry, the clothes can be hung outside and will dry naturally, or a well ventilated indoor space using sunshine or internal heat should suffice.

Electrical utility rate structures may have a bearing on the utilization of an appliance and at what time during the day it should be used. Some companies charge a lowered rate for electricity consumed during off-peak periods and tailor their rate structure to the maximum amount of electricity used at any one time. Local utility companies should be consulted to find out their regulations for determining rates.

It may be desirable to take advantage of off-peak rates, while keeping the usage at any one time to a minimum. This would mean that certain appliances may have to be used in the early morning or late evening hours. At times, appliances may have to be switched off to keep total usage to a minimum.

One potential use of off-peak electricity is as an auxiliary heating source in solar heated buildings. The heat must be stored so that it is available during peak hours. This can be accomplished by lacing an insulated gravel storage bin with heating coils, which heat the gravel during off-peak hours. When heat is required, air can be drawn through the bin where it will absorb heat (or a series of electric hot water tanks can be phased to receive off peak electric heat), and then deliver it to the building.

The wise use of appliances is necessary for energy conservation in homes and buildings. Appliances not only consume energy directly, but may contribute to increased cooling loads and further energy consumption. Below is a list of suggestions for the energy wise use of appliances.

1. Turn off all appliances (television, radio, phonograph, lights, etc.) when they are not being used.

2. Make sure all appliances are operating at maximum efficiency by having them serviced periodically.

3. Check refrigerator and freezer door seals to insure that they are air tight and that their condensing coils are clean, allowing good air flow. Open the doors as little as possible.

4. Operate heat producing appliances in the early morning or late evening when air temperatures are cooler and heavy demands are not being made upon electrical distribution systems.

5. "Instant on" television sets use energy even when the screen is dark. This waste can be eliminated by plugging the set into an outlet that is controlled by a wall switch and turning the set on and off with the switch, or install an additional on-off switch on the set itself or in the power cord.

6. Check the Energy Efficient Ratios (EER's) of appliances before buying. The higher the EER the more efficient the appliance. All room air-conditioners are now rated with this system, and the Department of Commerce is working to label other types of appliances.

7. Match appliance size to the size of the job, especially major appliances. Instead of cooking one potato or meat dish, cook an entire oven meal. For small meals a "toaster oven" may be most efficient.

8. Run dishwasher and clothes washer with full loads, but do not overload.

9. Use flat bottom pans that completely cover the burner or heating element of the stove. More heat will enter the pot and less will be lost to the surrounding air.

10. Maintain electrical tools in top operating shape. Keep them clean, sharp, and properly lubricated.

11. When purchasing a water heater, match its size to the needs it serves. Oversized water heaters use more energy than necessary.

12. Use a small appliance in place of a major appliance whenever possible.

13. Do not preheat appliances longer than necessary.

14. Don't overload any electrical circuits, as this results in reduced energy efficiency.

15. Use the waste heat from major appliances, other equipment, and lighting fixtures to assist with cold weather space heating. Vent this heat to the exterior during warm or hot weather.

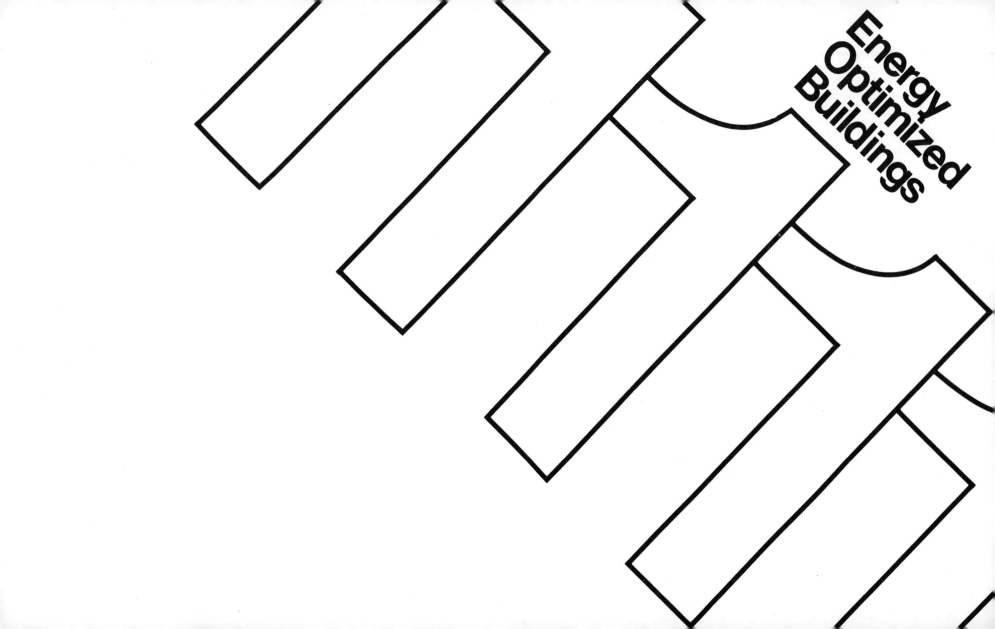

Energy
Optimized
Buildings

In this chapter a number of case studies of energy optimized buildings are presented. They have all been designed to optimize the conservation of energy through architectural design and site planning, to optimize the use of cost effective natural climatic systems, and to integrate the active solar collection and thermal storage to architecture and interior design.

Principal conservation features are indicated on the drawings, and many other elements which are integrated into the architectural system are presented in the text. However, each building may have features which are not reported. The primary emphasis of presentation is on the individuality of systems, and important design features which make these buildings both functional and practical.

The first four houses are located in a well established residential neighborhood, close to the Cherry Creek shopping area east of downtown Denver. Each of these buildings is unique and has certain features which are different from the others. This was done so that optimum designs, orientations, and configurations could be determined under actual service conditions. In addition, it clearly demonstrates that buildings employing energy conservation and solar collectors do not all have to look alike.

The first two buildings are situated on the same city lot. They are designated as "Cherry Creek Residence – 419A" and "Cherry Creek Residence – 419B".

The "419A" building is a 2000 square foot residence designed on a 28' x 28' foundation. It is intended as a prototype for either on-site or modular construction for townhouses, cluster housing, or single family use. The room arrangement is a square stacking plan with bedrooms and greenhouse on the lower level; living, dining and kitchen on the main level; and studio at the third level. Full bathrooms, including shower, are located on the lower and main levels.

The primary entry faces east away from north and northwest winter winds and is an airlock, doubled door, unheated vestibule. Glass area is limited, yet provides visual openness, attractive outdoor vistas, and natural illumination. Walls and ceilings are shaped and painted to function as light fixtures without glare. All windows are fabricated from clear plate, double glazed, insulating glass. The window position and size is calculated to receive direct winter solar gains but exclude them in the summer.

The hood over the kitchen range recirculates the air through a charcoal filter for cleaning and deodorization. The bathrooms have recirculative fans with citrus base filters that deodorize the air and kill bacteria. These systems prevent direct internal heat losses that occur through externally vented exhaust fan systems.

During spring, summer and fall, ventilation is provided by means of a roof exhaust turbine. Fresh air is provided in winter by the solar heated greenhouse and from return air that passes through an electrostatic and charcoal filtration system.

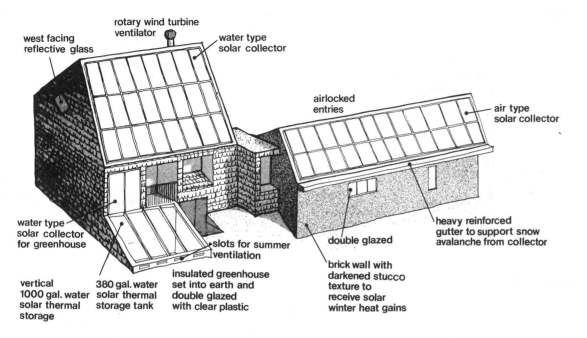

west facing reflective glass

rotary wind turbine ventilator

water type solar collector

airlocked entries

air type solar collector

water type solar collector for greenhouse

slots for summer ventilation

double glazed

heavy reinforced gutter to support snow avalanche from collector

vertical 1000 gal. water solar thermal storage

380 gal. water solar thermal storage tank

insulated greenhouse set into earth and double glazed with clear plastic

brick wall with darkened stucco texture to receive solar winter heat gains

data

CHERRY CREEK RESIDENCE "419A"

Floor Area	2000 square feet
Collector Area	560 square feet
Solar Collector	Liquid type (water)
Solar Collector Supplier	R-M Products
Heat Storage	1000 gallon water tank
Completed	1974
Architect	Richard L. Crowther AIA

data

CHERRY CREEK RESIDENCE "419B"

Floor Area	1100 square feet
	(+900 square foot basement)
Collector Area	384 square feet
Solar Collector	Air type
Solar Collector Supplier	Solaron Corporation
Heat Storage	30 tons of gravel
Completed	1974
Architect	Richard L. Crowther AIA

Outer walls of the house are ten inch wood studs with 9½" of rockwool insulation, an outer 3/4" thick insulating sheathing board, 5/8" of wood ply sheathing, and wood shakes to trap air. Internal 4 mil pliofilm acts as a continuous vapor barrier on the entire interior and is covered with ½" interior drywall. The "U" factor of this wall is less than .025, and due to its thickness provides a higher thermal mass, for retention of interior energy, than conventional frame construction. The roof construction is similar to that of the walls.

The 560 square foot solar collector, supplied by R-M Products of Denver, is a flat plate water type south facing roof collector set at a 53° angle. Each double glass cover plate is 1/8" thick. The interior coating is flat black. A reinforced gutter and top member are provided to support a movable ladder for collector access and to act as a snow trap. Heated water is piped from the rooftop solar collector to a 1000 gallon vertical fiberglass insulated storage tank. At night and when freezing conditions exist, the water is automatically drained from the collector into the insulated tank. This prevents damage to the collector resulting from the expansion of water upon freezing.

Cooling is provided on hot days by a 5400 cfm evaporative cooler, combined with the roof turbine fan exhaust system. The solar collection system is used to heat domestic water. Over 95% of all hot water is provided in this way. It is projected that 96% of all heating requirements will be satisfied by the solar collection system in conjunction

with the energy optimized design. In actual use,
90% of all cooling requirements were met with the
above mentioned system.

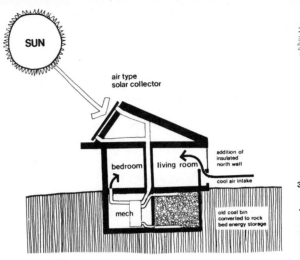

SUN

air type
solar collector

addition of
insulated
north wall

bedroom   living room

cool air intake

mech   old coal bin
converted to rock
bed energy storage

419 B

SUN

high return air

electrostatic
and charcoal
filters

furnace

recirculating
rush-hampton
CA 90 fan

greenhouse
solar collector   living room   bath

380 gal. hot water storage
greenhouse

fresh air intake   bedroom   bath

419 A

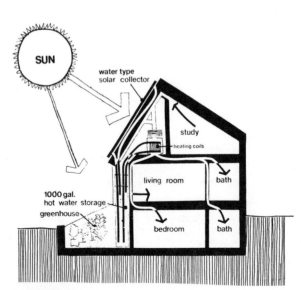

SUN

water type
solar collector

study

heating coils

living room   bath

1000 gal.
hot water storage

greenhouse

bedroom   bath

419 A

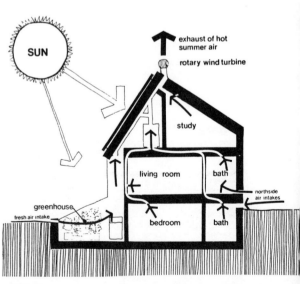

SUN

exhaust of hot
summer air

rotary wind turbine

study

living room   bath

northside
air intakes

greenhouse

fresh air intake   bedroom   bath

419 A

The "419B" building is the result of extensive remodeling of an older 1100 square foot residence. It also has a 900 square foot basement. To improve the total thermal characteristics of the house, the basement is utilized as a thermal reservoir, to moderate the heating and cooling needs of the house by means of direct transfer grilled floor openings. The old coal bin is re-used for solar thermal storage. It is fitted with a layer of open concrete block covered with wire mesh to serve as a lower manifold, and filled with 20 cubic yards of large gravel. A new insulated north wall was constructed the length of the house that shields the north side from winter storms and winds, as well as converting the old uninsulated 8" brick wall to an interior thermal inertia wall. The new north wall extends out from the old wall approximately 3', allowing for moderate day and cool nocturnal temperatures to be drawn in through recessed, insulated air vents. The south roof angle was altered to 53° for effective use of an air type solar flat plate collector. Six inches of insulation was added to the roof to supplement the four inches of existing loose fill insulation.

The main entry faces east and is an airlock double door unheated vestibule. A floor grille to the basement near the entry picks up cold air intrusion. The old front entry door is replaced with a new weatherstripped door, and the old porch converted to a solar patio. Small, old northside windows are replaced with insulated weatherstripped solid panels which can be opened for natural ventilation. On the south side of the living room, a new slimshade double glazed win-

dow has been provided and is protected by a roof overhang. Two east windows with sun control, heat absorbing gray double glazed windows extend to the floor and replace the old double hung windows. The west window in the bathroom was replaced with fixed double glazed reflective glass, to eliminate outside air infiltration and decrease west solar radiation approximately 78%.

Bathroom and kitchen fans are the recirculating type to decrease heat losses. The exterior of the house was completely refinished with cedar wood shakes on the roof and sidewalls, adding extra insulation to them. Other sidewalls are finished with new stucco.

The 384 square foot solar collector, supplied by Solaron Corporation of Denver, is a flat plate air type collector. Each double glass cover plate is 1/8" thick. The interior coating is flat black. A reinforced gutter and top member is provided to support a movable ladder for collector access and to act as a snow trap. Solar heated air is forced through a duct system to the thermal gravel storage bin in the basement.

During daytime periods of full or diffuse sunlight, solar heated air travels through the gravel storage bin and the existing furnace before being delivered to the house. During overcast periods or at night, the forced air system will draw solar heated air from the gravel storage bin. Natural gas is the back-up fuel source. During seasons when the house needs cooling, cool daytime and nocturnal air will cool the gravel bin, and cool air will be delivered as needed through the house by means of the existing forced air system.

It is projected that 85% of all heating requirements will be provided by the solar collection system in conjunction with energy optimized features. Cooling is provided by natural ventilation.

The third building in this group is "Cherry Creek Residence - 435". This unit, on a 20' x 20' foundation, is designed as a prototype for single family, duplex or cluster housing. It is a two story, low cost, 800 square foot home with one bedroom, small den, greenhouse, and a combination kitchen, dining and living room. In addition it has a 200 square foot attic storage area behind the solar collector, which is accessible by a pulldown ladder stairway.

The 240 square foot solar collector, supplied by Solaron Corporation, is a flat plate air type south facing roof collector, set at 50° to optimize winter solar radiation. Delivery of the solar heated air is either made directly to the house, or when internal heating is satisfied it is stored in an insulated gravel bin. The storage bin is located under the house and does not take up interior space. Auxiliary heating is provided by separate radiant electric base heaters located in each room. The temperatures of each room, maintained by the auxiliary system, are individually controlled by remote thermostats.

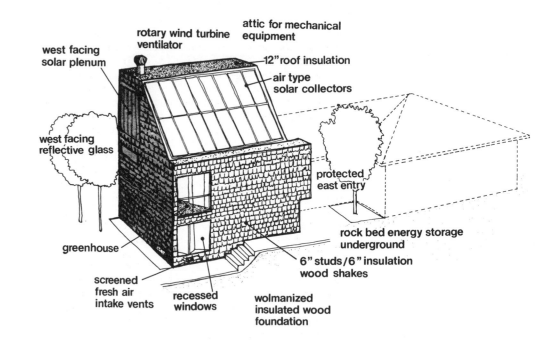

rotary wind turbine ventilator

attic for mechanical equipment

west facing solar plenum

12" roof insulation

air type solar collectors

west facing reflective glass

protected east entry

rock bed energy storage underground

greenhouse

6" studs/6" insulation wood shakes

screened fresh air intake vents

recessed windows

wolmanized insulated wood foundation

data

CHERRY CREEK RESIDENCE "435"

Floor Area	800 square feet
	(+200 square foot attic)
Collector Area	240 square feet
Solar Collector	Air type
Solar Collector Supplier	Solaron Corporation
Heat Storage	12 tons of gravel
Completed	1975
Architect	Richard L. Crowther AIA

A west facing solar plenum with built-in thermal
storage capacity provides inductive ventilation.
The house is also equipped with a rotary wind
turbine ventilator. Individual vents are located
near the lower level floor line to allow the entry of
cool nocturnal or cool daytime air. These reflective
metal ventilation units have weather hoods, avoid
direct wind flow, and have thick insulative doors
with gaskets to prevent infiltration when closed.

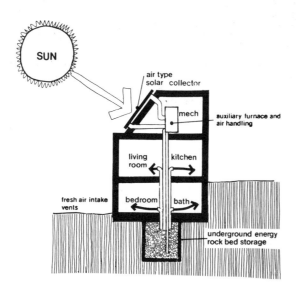

SUN

air type
solar collector

mech

auxiliary furnace and
air handling

living
room

kitchen

fresh air intake
vents

bedroom

bath

underground energy
rock bed storage

The building envelope is minimized by the square
configuration, low ceiling heights (7'4"), and by
setting portions of the lower level below the grade
line. External walls are constructed of specially
treated 6" wood studs continuous from below grade
foundations to the ceiling line of the bedrooms.
Regular 6" wood studs continue to the roof line.
Handsplit heavy red cedar wood shakes over water-
proofed felt and plywood sheathing with taped joints
to avoid air intrusion form a durable, no maintenance, energy conserving barrier. Six inches of
rockwool batts in outer stud walls are used continuously from 3' below grade to the roof line.
The roof areas have 12" of rockwool batts. These
batts are backed with plastic vapor barriers to the
interior of the house.

Total glass area to the exterior is approximately
10% of the total floor area. All windows are fixed
insulating units set in a glazing bed to avoid infiltration and reduce thermal losses.

If produced in quantity with organized field construction methods, this home could take the place
of mobile homes. This prototype was constructed
in five weeks.

Optimized energy conservation and passive systems
provide 70% of the actual cooling of this unit. Pro-
jections show that 80% of the heating will be provided
by the solar collection system in conjunction with the
other optimization features.

The last building in this area is "Cherry Creek Residence – 500". It is a 2600 square foot residence with a 1250 square foot apartment, designed to optimize energy conservation and utilize passive solar collection. The entire building is designed to function as a solar collector in the winter and prevent solar gains in the summer. Heat storage is provided by the thermal inertia of floors and walls.

The principal entry for each unit is recessed to protect it from prevailing winter winds. All windows are double glazed, fixed glass. Window areas are kept to a minimum, yet their shape and location maximize interior daylight. Natural illumination is also provided by double plastic, heat absorbing skydomes and special skyshafts with near zero thermal loss. Artificial lighting levels are provided on a task basis.

Solar gain through southern exposure is maximized during winter and minimized during summer, by calculated window recess depths and outdoor deciduous vegetation combinations. The southern windows of the dining room function as two-stage winter solar collectors, with the lower unit active from September 1 until April 1 and the upper unit active from October 15 to February 15.

The exterior wall color was selected to balance radiant heat gains and losses on a 12 month basis. The rough exterior wall texture deepens the stationary air film, as a thermal buffer zone. The "U" factor of the walls, roof, and floor is .05.

Garage and storage area locations provide thermal buffering of internal living spaces. Modulated ceiling

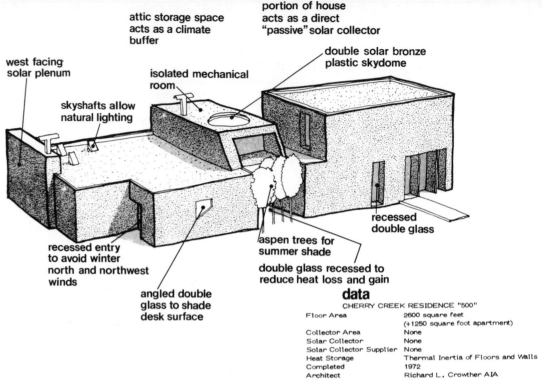

attic storage space acts as a climate buffer

portion of house acts as a direct "passive" solar collector

west facing solar plenum

isolated mechanical room

double solar bronze plastic skydome

skyshafts allow natural lighting

recessed double glass

recessed entry to avoid winter north and northwest winds

aspen trees for summer shade

double glass recessed to reduce heat loss and gain

angled double glass to shade desk surface

data
CHERRY CREEK RESIDENCE "500"

Floor Area	2600 square feet
	(+1250 square foot apartment)
Collector Area	None
Solar Collector	None
Solar Collector Supplier	None
Heat Storage	Thermal Inertia of Floors and Walls
Completed	1972
Architect	Richard L. Crowther AIA

heights are used to direct the hottest air to the highest point in the building, allowing interior air temperatures to be salvaged by high-level air returns in winter and low-level air returns in summer. For summer cooling roof plenums, heated by the sun, induce upward air motion which allows outside air to enter by displacement through north vents.

The passive systems and energy optimized features of this home provide 65% of the heating and 60% of the cooling. The house is also equipped with conventional heating and air conditioning units. These

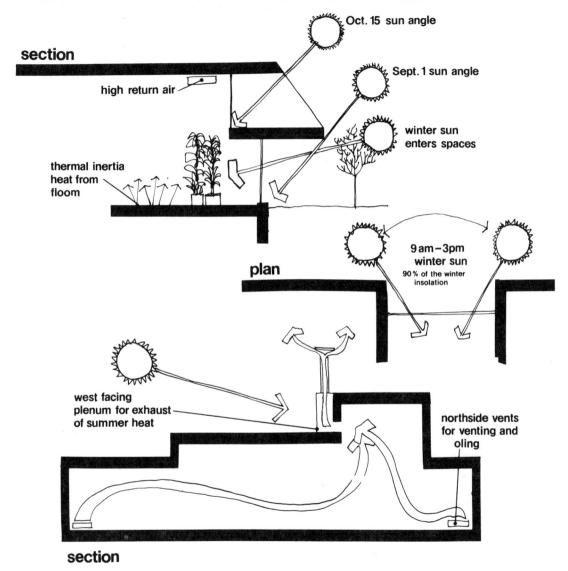

section

Oct. 15 sun angle

high return air

Sept. 1 sun angle

thermal inertia
heat from
floom

winter sun
enters spaces

plan

9 am – 3pm
winter sun
90% of the winter
insolation

west facing
plenum for exhaust
of summer heat

northside vents
for venting and
oling

section

systems contain electrostatic, charcoal, and fiber filters which allow interior air to be recirculated without the intrusion of exterior air which may be at a different temperature, thereby salvaging thermal equilibrium.

Large interior plants provide oxygen to the air and absorb carbon dioxide. Winter humidification of interior air is produced by a water fountain, which raises the perceived temperature.

The "CSU Solar House I" was built by the Solar Energy Applications Laboratory of Colorado State University, using a grant from the National Science Foundation. It was the first residential-size building in the world to be actively heated and cooled with solar energy. The 3000 square foot building was designed as a solar prototype and demonstration project.

The 768 square foot flat plate liquid type solar collector is located on the southern roof area and is set at a 45° angle. This orientation of the collector favors the optimum collection of solar energy during the entire year.

Heat which is absorbed by the metal surface of the collector is transferred to the collector fluid, a 25% solution of ethylene glycol in water. In order to minimize the amount of ethylene glycol needed, the collector fluid does not pass directly into the storage tank but instead is piped through a heat exchanger. The heat is transferred to water, and the water is piped to storage.

Architectural energy conservation features of this house include vertical exterior fins to reduce convective losses from window and wall areas, double glazed wood sash windows, recessed airlock entry, garage positioned as a climate buffer, and portions of the structure set into the earth. Overhangs provide summer shading of windows, yet allow the winter sun to penetrate.

Operation, monitoring, and evaluation of the CSU Solar House I is under the direction of Dr. George O.G. Lof. Solar energy is used in this house for space heating, domestic hot water heating, and to power a lithium bromide absorption cooler. Heating and cooling are forced air systems. Solar energy actually provides 83% of the heating and 43% of the cooling, as well as most of the domestic hot water requirements. With modifications, solar energy is expected to provide 79% of the cooling.

Colorado State University has constructed two additional solar laboratory buildings to be equipped with heating and cooling systems of different designs. Solar House II uses air to collect solar energy in the form of heat, while Solar House III uses a liquid. With these three houses, it will be possible to operate three different systems at the same time, under the "same" sun and weather conditions. With this arrangement, direct comparison of the performance of the three heating and cooling systems is possible.

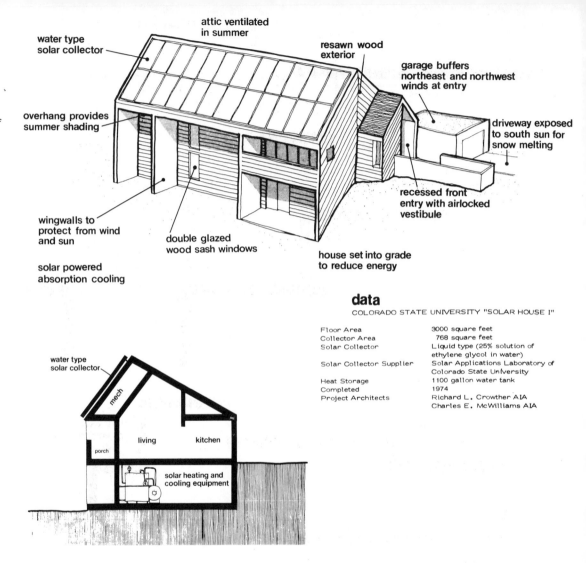

water type solar collector

attic ventilated in summer

resawn wood exterior

garage buffers northeast and northwest winds at entry

overhang provides summer shading

driveway exposed to south sun for snow melting

recessed front entry with airlocked vestibule

wingwalls to protect from wind and sun

double glazed wood sash windows

house set into grade to reduce energy

solar powered absorption cooling

water type solar collector

mech

porch

living

kitchen

solar heating and cooling equipment

data
COLORADO STATE UNIVERSITY "SOLAR HOUSE I"

Floor Area	3000 square feet
Collector Area	768 square feet
Solar Collector	Liquid type (25% solution of ethylene glycol in water)
Solar Collector Supplier	Solar Applications Laboratory of Colorado State University
Heat Storage	1100 gallon water tank
Completed	1974
Project Architects	Richard L. Crowther AIA
	Charles E. McWilliams AIA

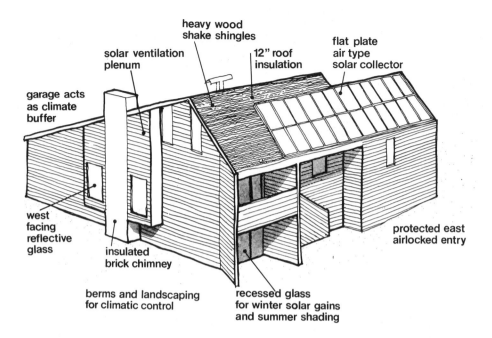

heavy wood shake shingles

solar ventilation plenum

12" roof insulation

flat plate air type solar collector

garage acts as climate buffer

west facing reflective glass

insulated brick chimney

berms and landscaping for climatic control

recessed glass for winter solar gains and summer shading

protected east airlocked entry

This suburban community residence of 3800 square feet on three levels is located in the pine hills twenty miles southeast of Denver, near Pinery, Colorado. A similar home is under construction at Stansbury Park, Utah. It has a 585 square foot air type solar collector, supplied by Solaron Corporation of Denver. Heat is stored in a bin located on the lower level.

The garage and earth berms are positioned to act as buffers to protect the home from heat loss due to cold northwest winter winds. Heavy wood shake shingles and 12" of insulation in the roof decrease the penetration of heat through this area. Double door airlock entries are used to minimize the intrusion of cold winter air and warm summer air.

Southern windows are double glazed and recessed for winter solar gains and summer shading. Mountain view west windows are glazed with reflective glass and are cooled by west facing solar ventilated plenums.

In the winter, heated air at the third level ceiling is returned to the flat plate solar collector and reheated for direct space use or storage. This air can be used by the collector for flow through summer ventilation and cooling of the home and the collectors. Additional air flow for venting of the collectors is provided by a fan.

It is projected that the energy optimization features and solar collection system will provide 80% of the space heating. These systems along with natural ventilation are projected to supply 50% of the house cooling.

data

SUBURBAN COMMUNITY RESIDENCE

Floor Area	3800 square feet
Collector Area	585 square feet
Solar Collector	Air type
Solar Collector Supplier	Solaron Corporation
Heat Storage	15 tons of gravel
Completed	Early 1976
Architect	Richard L. Crowther AIA
	Crowther/Solar Group

These two suburban solar homes, designed as prototypes for a housing development west of Denver, feature a two story, 2100 square foot home, and a one story, 1000 square foot house. Both have Solaron Corporation air type solar collectors and gravel heat storage. The two story unit has 351 square feet of collector area and 9 tons of gravel storage, while the other unit has 234 square feet of collector area and 6 tons of gravel storage.

Architectural features of the two story house are minimal northside windows to decrease heat loss, wingwalls to protect westside windows from wind and sun, and a recessed south entry. Excess greenhouse heat is ventilated through a roof stack in the summer, and used for space heating during the winter.

The smaller house has no west side windows and minimal northside glazing. It has a south facing entry and patio, protected from direct afternoon sun by a wingwall. The heat storage area is located directly behind the solar collector. Both houses have 4" wood stud walls, insulated with moisture barrier batt type insulation. Dark colored wood siding increases heat gains when the sun angle is low (winter) and the light color roof reflects heat when the sun angle is high (summer). Double glazed, wood frame windows are used throughout.

These houses were designed to be economically competitive with other development homes. The solar heating systems are expected to provide 75% of all space heating. Solar heating of domestic hot water is being considered.

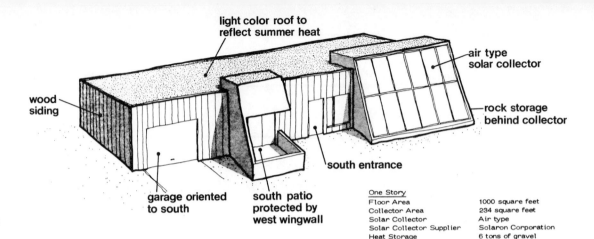

light color roof to reflect summer heat

air type solar collector

wood siding

rock storage behind collector

south entrance

garage oriented to south

south patio protected by west wingwall

One Story
Floor Area	1000 square feet
Collector Area	234 square feet
Solar Collector	Air type
Solar Collector Supplier	Solaron Corporation
Heat Storage	6 tons of gravel
Completed	Undetermined
Architect	Richard L. Crowther
	Crowther/Solar Group

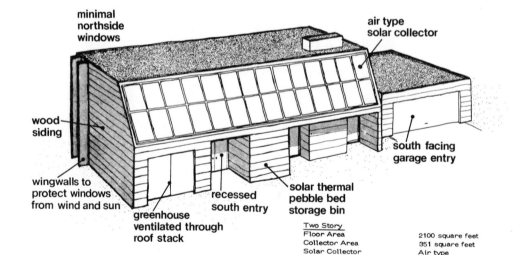

minimal northside windows

air type solar collector

wood siding

south facing garage entry

wingwalls to protect windows from wind and sun

recessed south entry

solar thermal pebble bed storage bin

greenhouse ventilated through roof stack

Two Story
Floor Area	2100 square feet
Collector Area	351 square feet
Solar Collector	Air type
Solar Collector Supplier	Solaron Corporation
Heat Storage	9 tons of gravel
Completed	In bidding phase
Architect	Richard L. Crowther AIA
	Crowther/Solar Group

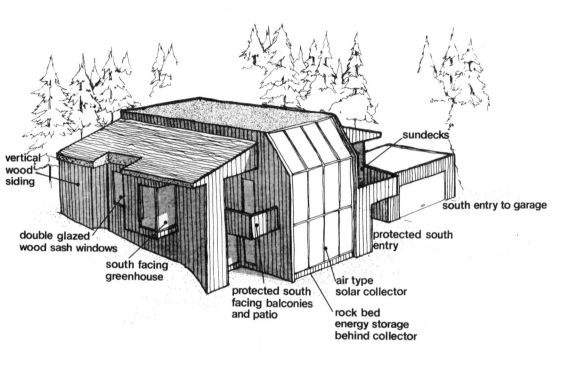

vertical wood siding

sundecks

double glazed wood sash windows

south entry to garage

south facing greenhouse

protected south entry

protected south facing balconies and patio

air type solar collector

rock bed energy storage behind collector

data

MOUNTAIN RESIDENCE NEAR GOLDEN, COLORADO

Floor Area	2500 square feet
Collector Area	370 square feet
Solar Collector	Air type
Solar Collector Supplier	Solaron Corporation
Heat Storage	7 tons of gravel
Completed	1975
Architect	Richard L. Crowther AIA

This 2500 square foot residence with a mountain setting is located in a pine forest clearing northwest of Golden, Colorado. It is designed to have optimal views in the mountain forest, yet has a total glass area of less than 10% of the total floor area. All glass is double pane "insulating glass" set in wooden frames.

The exterior siding is natural exposed wood. The house is partially set into the mountainside to reduce energy losses. The four bedroom area is located on the lower level, with the primary living areas on the upper level. This allows for sleeping areas to be cooler than living areas. The plan arrangement provides interior buffering elements on the north side to protect from prevailing winter winds. Ceiling heights, room volumes, and openings are designed to accumulate and combine internal energy with solar energy.

The air type solar collector was supplied by Solaron Corporation of Denver. It has 370 square feet of area and supplies heat to seven tons of gravel storage. Portions of the collector are both tilted and vertical. The vertical collector will provide about 10% less energy during the heating season than one tilted to the optimum heating orientation. The solar heating system is also used to heat domestic hot water. It is projected that 80% of the space heating requirements will be met by the solar heating system, along with a large portion of the domestic hot water. Cooling is provided by natural ventilation.

This 3600 square foot house is located in the mountains west of Denver. The requirement for an interior greenhouse and large south windows necessitated installing movable insulated panels. These panels are open for approximately 8 hours a day during the winter, then are closed to protect against severe nighttime heat loss. During the summer, the panels are closed during the day and opened at night. Natural illumination is provided at all times by clerestory windows which consist of fiberglass panels sandwiched over translucent insulation. These windows are located above the greenhouse and second floor areas.

Heating equipment includes 576 square feet of air type solar collectors supplied by Solaron Corporation of Denver, and 11 tons of gravel for heat storage. Space heating and domestic hot water are provided by this system. Auxiliary heat is provided through a second gravel storage bin. The gravel in this second bin is heated by resistance coils. When air is drawn through it, it functions as an electric forced air furnace. The electric coils are in operation only during off-peak periods. It is designed to be able to take advantage of electrical rate structures which favor off-peak usage.

The solar heating system is projected to supply 75% of annual space heating and up to 85% of domestic hot water. Thermal gains during the summer months are vented through two roof turbines located nearly thirty feet above the ground floor. This is a sufficient height to develop strong stack action, which is able to ventilate large volumes of air. The roof vents contain motorized insulated dampers.

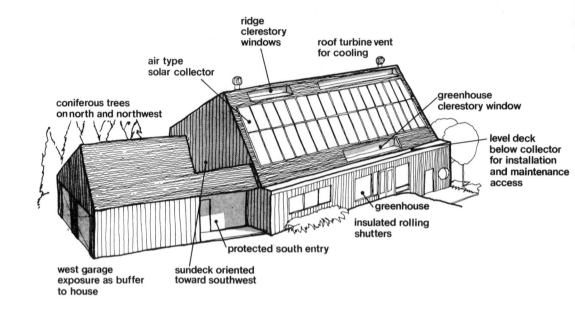

ridge
clerestory
windows

roof turbine vent
for cooling

air type
solar collector

coniferous trees
on north and northwest

greenhouse
clerestory window

level deck
below collector
for installation
and maintenance
access

greenhouse

insulated rolling
shutters

protected south entry

west garage
exposure as buffer
to house

sundeck oriented
toward southwest

data

MOUNTAIN RESIDENCE WEST OF DENVER

Floor Area	3600 square feet
Collector Area	576 square feet
Solar Collector	Air type
Solar Collector Supplier	Solaron Corporation
Heat Storage	11 tons of gravel
Completed	Summer 1976
Architect	Lawrence C. Atkinson
	Crowther/Solar Group

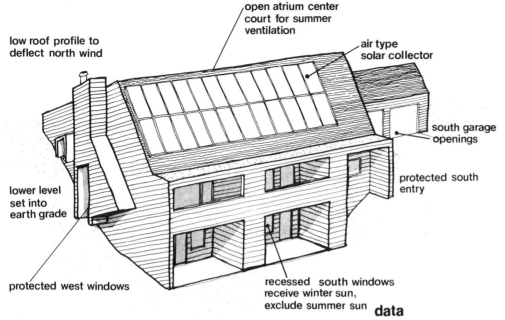

open atrium center court for summer ventilation

low roof profile to deflect north wind

air type solar collector

south garage openings

lower level set into earth grade

protected south entry

protected west windows

recessed south windows receive winter sun, exclude summer sun

data

SUBURBAN WYOMING RESIDENCE

Floor Area	3200 square feet
Collector Area	460 square feet
Solar Collector	Air type
Solar Collector Supplier	Solaron Corporation
Heat Storage	15 tons of gravel
Completed	Spring 1976
Designer	Paul Karius
	Crowther/Solar Group

This 3200 square foot residence is located near Cheyenne, Wyoming. The high winter winds and sloping site made special design features necessary. The shape of the house minimizes wind turbulence and its north side is set into the hillside, both minimizing the amount of heat lost to prevailing winter winds. The windows on the west side are protected by overhangs and sidewalls, to decrease convective heat losses from the glass. All windows are double glazed and set in wooden frames.

The house was located on the south side of the site, so that the slope there could be used to maximize the southern exposure of the house, while the northern exposure was minimized. The windows on the south are recessed to allow for summer shading and winter thermal gains. The northern area of the site is flat, and in the future a wind machine may be installed there.

The house is provided with a protected southeast airlocked entry. All walls are constructed with 2" x 6" studs, and have 5½" of R-19 foil backed batt insulation. The roof is constructed with 2" x 10" joists, and has 9½" of R-30 foil backed batt insulation. The foil forms a vapor barrier around the building envelope. Basement walls are insulated with 2" of styrofoam.

The heat from the 460 square foot air type solar collector, supplied by Solaron Corporation of Denver, is stored in 15 tons of gravel.

It is projected that 80% of the annual heating requirements will be provided by solar collection and optimized energy conservation. A large percentage of the domestic hot water will also be supplied by the solar collection system. Natural ventilation will be used for cooling and will be induced by the solar collector with its fan. Hot air, which collects at the highest point in the house, is piped to the solar collector and is heated further. It is then routed to a heat exchanger for hot water heating before being vented to the outside. One hundred percent of summer hot water needs are expected to be provided with this system.

187

This 2000 square foot residence is built on farm-land northwest of Denver. Summer heat gains through south windows are virtually eliminated by their recess depth, but winter sun penetrates them and helps warm the interior.

The house is equipped with a 385 square foot solar collector, supplied by Solaron Corporation of Denver. Heat is stored in 12 tons of gravel. Roof turbine ventilators exhaust hot summer air, drawing in cool air through low-level vents to provide non-mechanical summer cooling. The roof is constructed of a dark material, to assist in the summer heating of a plenum for increased stack action effect. This also helps to add winter heat to the building, primarily the garage. The garage door is also painted dark for the purpose of winter heating.

The north side has minimal window area to protect against heat loss to prevailing winter winds. West side windows are reflective so that summer after-noon heat gains are reduced. The wood siding is a lighter color than the roof and was selected to balance year-round heat gains through the building envelope. The east-west dimension of the overall house is 1.6 times the north-south dimension, which optimizes the annual heat gain/heat loss ratio for this climatic region.

The solar collection system in conjunction with the optimized energy conservation features of this house is expected to supply over 80% of the annual heating requirements. Natural ventilation will provide a large portion of the cooling.

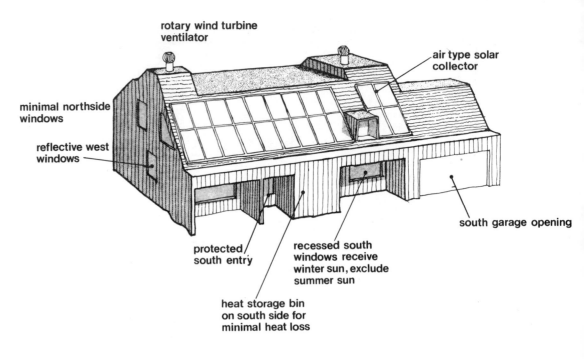

rotary wind turbine ventilator

air type solar collector

minimal northside windows

reflective west windows

protected south entry

recessed south windows receive winter sun, exclude summer sun

heat storage bin on south side for minimal heat loss

south garage opening

data

FARMLAND RESIDENCE

Floor Area	2000 square feet
Collector Area	385 square feet
Solar Collector	Air type
Solar Collector Supplier	Solaron Corporation
Heat Storage	12 tons of gravel
Completed	Spring 1976
Architect	Lawrence C. Atkinson Crowther/Solar Group

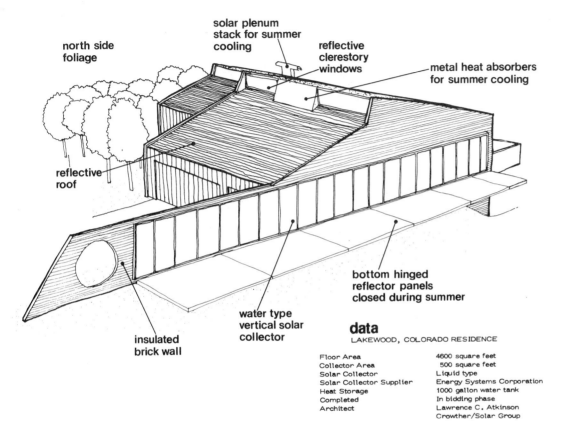

north side foliage

solar plenum stack for summer cooling

reflective clerestory windows

metal heat absorbers for summer cooling

reflective roof

bottom hinged reflector panels closed during summer

insulated brick wall

water type vertical solar collector

data
LAKEWOOD, COLORADO RESIDENCE

Floor Area	4600 square feet
Collector Area	500 square feet
Solar Collector	Liquid type
Solar Collector Supplier	Energy Systems Corporation
Heat Storage	1000 gallon water tank
Completed	In bidding phase
Architect	Lawrence C. Atkinson
	Crowther/Solar Group

This 4600 square foot residence will be located in the city of Lakewood near Denver. The 500 square foot liquid type solar collector, supplied by Energy Systems Corporation of Colorado Springs, will provide this home with space heating and domestic

hot water. When the temperature of the water in the collector is less than the temperature of the water in storage, the water in the collector will automatically drain into the 1000 gallon storage tank.

The home is set into a hillside for thermal protection and opens onto an outdoor patio and tennis court. Interior volumes and room orientation are planned for maximum internal energy reuse. Hinged reflective panels are positioned in front of the vertical solar collector. During the winter when they are open, the collector receives reflected, as well as direct radiation. During the hottest summer periods the reflective panels are closed against all the collector modules except the ones providing domestic hot water. These modules are separated from the others by a room window, which in turn is obscured from exterior view by tinted glass which matches adjacent collector modules.

Reflective clerestory windows repel up to 80% of the afternoon heat gains from the intense west sun, while allowing ample natural illumination to enter the building. Sheetmetal panels between the two clerestory windows are used during the summer to induce the upward flow of air by stack action. Cool ground level air enters by displacement through specially designed vents.

It is projected that 75% of the heating and a large percentage of the domestic hot water will be provided with solar energy and optimized energy conservation. The auxiliary heating system is a gas fired furnace which delivers its heat by forced air through the same ductwork as the solar heat. Cooling is provided by natural ventilation.

The mountainous region near Gunnison, Colorado, forms a beautiful setting for this 2500 square foot residence. The house is supported on wooden foundations, chemically treated with chromated arsenate for permanence. Wall and foundation construction of 6" wood studs provides space for 6" of fiberglass batt insulation (R-19) to be continuous from the wooden foundation plate to the roof line.

The 600 square foot air type solar collector, supplied by Solaron Corporation of Denver, provides space heating and domestic water preheating. Heat is stored for distribution in a well insulated bin filled with 15 tons of gravel. A double-dampered fireplace and an auxiliary system of radiant electric baseboard heaters supply the remainder of the heating needs. The fireplace is equipped with a system of low-level cool air intake vents, ducts running next to the firebox in which the air is heated, and a hot air outlet which returns the air to the room. This system salvages some of the heat that would normally be lost up the flue.

The solar collection system in conjunction with optimized energy conservation features supplies this house with 85% of its annual space heating and a large percentage of its hot water. Cooling is accomplished by natural ventilation.

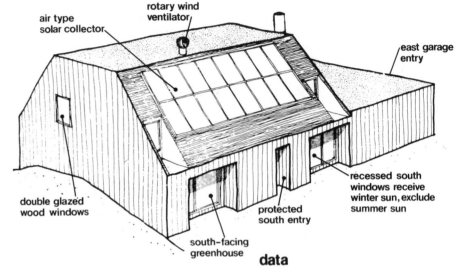

air type solar collector

rotary wind ventilator

east garage entry

double glazed wood windows

south-facing greenhouse

protected south entry

recessed south windows receive winter sun, exclude summer sun

data

GUNNISON RESIDENCE

Floor Area	2500 square feet
Collector Area	600 square feet
Solar Collector	Air type
Solar Collector Supplier	Solaron Corporation
Heat Storage	15 tons of gravel
Completed	Spring 1976
Architect	Richard L. Crowther AIA
	Crowther/Solar Group

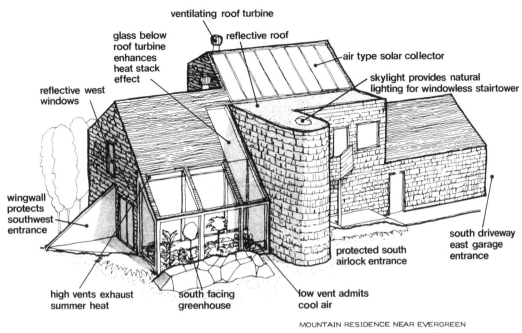

ventilating roof turbine

glass below roof turbine enhances heat stack effect

reflective roof

reflective west windows

air type solar collector

skylight provides natural lighting for windowless stairtower

wingwall protects southwest entrance

high vents exhaust summer heat

south facing greenhouse

low vent admits cool air

protected south airlock entrance

south driveway east garage entrance

MOUNTAIN RESIDENCE NEAR EVERGREEN

Floor Area	2364 square feet
Collector Area	480 square feet
Solar Collector	Air type
Solar Collector Supplier	Solaron Corporation
Heat Storage	9½ tons of gravel
Completed	In bidding phase
Architect	Lawrence C. Atkinson
	Crowther/Solar Group

The mountains west of Denver near Evergreen provide a majestic setting for this 2364 square foot home. It was designed to complement its location and was constructed with minimum disturbance to neighboring trees, rocks, and natural vegetation. Heat loss to northwest winter winds was minimized by limiting the

window area on the north and west sides, providing a wingwall on the west balcony to shield the sliding glass door, and arranging interior spaces to provide northern buffer areas. Heat from the sun is provided through southern windows, by a greenhouse, and an air type solar collector.

The kitchen, laundry area, mechanical room, closets, and rock bin for heat storage are located along the north wall, as buffer zones from cold winter winds. The storage bin itself is a narrow vertical chamber 2½ stories high and is located directly under the solar collector. Heat escaping from storage can be recaptured and routed back into the storage bin.

The greenhouse on the south side helps to warm the house in the winter and cool it in the summer. During winter days, the sun's energy enters through the glass and is transferred by convection to other rooms. In the summer, the heat from the greenhouse is used to induce ventilation which is further accelerated through the rotary roof turbine. The hot air is exhausted through the turbine and cool air enters through low-level vents to take its place.

It is projected that the solar collection system in conjunction with optimized energy conservation and climate responsive architecture will provide 80% of the annual space heating and 60% of the hot water for this building. Natural ventilation is the only system provided for cooling and is expected to provide comfortable temperatures throughout the hottest months. Auxiliary heat is produced and distributed by a gas fueled forced air system.

These energy optimized solar office buildings of unified design are located in the Cherry Creek area of Denver. Each one has 4500 square feet of floor area on two levels. The levels are divided by a central foyer to produce space which is suitable for use by one to four tenants.

Architectural energy conservation features are similar for both buildings and are described below.

The primary building entry faces south for winter sun exposure and as a protection from cold winter winds. The total glass area is limited to less than 10% of the total floor area, yet the size, location, and shape of windows provides visual openness and natural illumination. Some windows are recessed so that they receive direct winter solar gains but are excluded from the hot summer sun. West facing windows are covered with a reflective surface and resist the penetration of 78% of incident summer radiation. All windows are double glazed with insulating glass to minimize thermal transfer.

Wooden foundations are treated with chromated arsenate for permanence. Wall construction type is continuous from the wooden foundation plate to the top of the roof parapet. This system permits uninterrupted insulation to be placed in the walls from below floor slabs to above the roof line. Heat loss is greatly reduced using this system. Outer walls are 6" wood studs with 6" of fiberglass batt insulation, an outer ½" thick plywood layer, 1½" of styrofoam insulation, and "Tri-lite" exterior finish. Internal 6 mil pliofilm provides a continuous vapor

barrier and is covered with ½" of interior drywall. The roof structure has 16" of high density fiberglass batt insulation.

Heat loss is reduced by having the lower level of the building set into grade. Earth berms are used to direct wind flow away from building surfaces, reducing the infiltration of automobile exhaust and dust from a nearby street.

The 175 square foot flat plate air type solar collector, supplied by Solaron Corporation, is south facing and is tilted at a 45° angle to the horizon. A skylight is located on the same plane as the collector, at the collector's upper edge. An upper mirrored overhang reflects winter sun directly into the skylight, for interior illumination and interior heat gains to the north side of the building.

Direct light enters the skylight through the winter months, but not during the summer months when the sun is at a high angle.

The roof of the building is surfaced with white marble chips. During the winter when the sun angle is low, diffuse radiation is reflected into the collector, increasing the amount of heat at the absorber. In the summer when the sun angle is high, a large portion of the diffuse radiation is reflected back to the sky. This scheme increases heat gains in the winter and decreases them in the summer. Exterior walls are colored darker than the roof to balance year-round heat gains and losses.

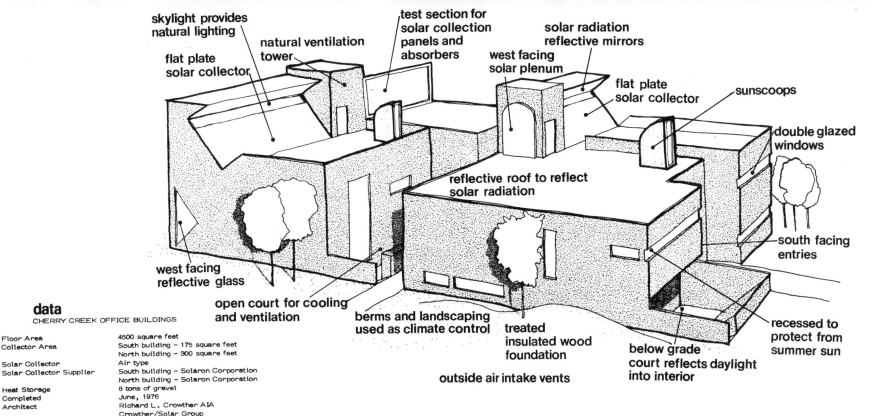

skylight provides natural lighting

test section for solar collection panels and absorbers

solar radiation reflective mirrors

flat plate solar collector

natural ventilation tower

west facing solar plenum

flat plate solar collector

sunscoops

double glazed windows

reflective roof to reflect solar radiation

south facing entries

west facing reflective glass

open court for cooling and ventilation

berms and landscaping used as climate control

treated insulated wood foundation

below grade court reflects daylight into interior

recessed to protect from summer sun

outside air intake vents

data

CHERRY CREEK OFFICE BUILDINGS

Floor Area — 4500 square feet
Collector Area — South building – 175 square feet
North building – 300 square feet

Solar Collector — Air type
Solar Collector Supplier — South building – Solaron Corporation
North building – Solaron Corporation

Heat Storage — 8 tons of gravel
Completed — June, 1976
Architect — Richard L. Crowther AIA
Crowther/Solar Group

The sun scoop over the entry stairway foyer has a south facing aperture glazed with double insulating glass. Inside the scoop behind the glass is a curved surface covered with reflective aluminum. Sunlight enters the glass, reflects from the aluminum, and is directed downward through an opening into the interior space for light and heating of the entryway. A duct with a fan on one end, reaching up inside the sun scoop, draws hot air down and delivers it to a storage bin filled with gravel located beneath the opening in the ceiling. Heat from the bin is radiated and convected to the entryway, thermally tempering this area.

Natural cooling is provided by inductive ventilation. Once the warm air reaches the highest point in the building, it enters ducts and is either vented by a wind turbine or is injected into the bottom of the solar collector and ventilates the collector before being exhausted by wind powered roof turbines. Cool nocturnal or late afternoon air is drawn into the building through ground level vents. These vents have weather hoods, prevent direct wind flow, and have insulated panels with gaskets to block infiltration when closed. Heat pumps are provided to remove heat from the interior when the outside temperatures are too high to be used for natural cooling.

Auxiliary heating is also provided by the rotary high performance heat pumps. One of their primary advantages is that they can be reversed. To cool the building they reject heat through roof mounted condensing units; to heat the building the roof condensers function as evaporators which extract heat from cold outside air, and supply it to the building through interior condensing units.

The building is divided into two heating and cooling zones, the temperatures of which can be controlled independently. One heat pump serves each zone, resulting in efficient energy usage. During heating periods, warm air that rises to the highest point in the building is captured and transported down to the lowest level. Before being reintroduced into the heating system, it is cleaned by fiber, charcoal, and electrostatic filters. Using this system, internal building heat is conserved and recycled.

In addition to the above features, the north building has a greenhouse on the lower level and a negative ion water fountain. The plants in the greenhouse will absorb carbon dioxide and produce humidity.

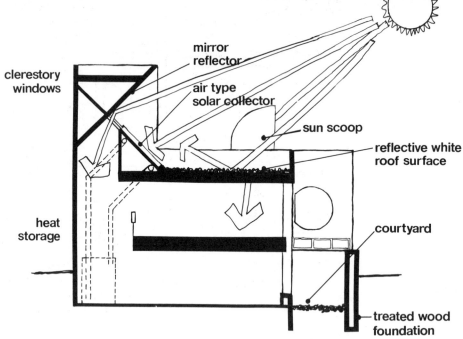

The unique feature of the south building is that it has a west facing heat box plenum. The plenum is warmed by the sun in the afternoon during spring, summer, and fall, and combined with the building's internal heat loads, increases inductive ventilation by producing stack action which is used to ventilate the building.

It is projected that solar energy, in conjunction with energy optimized architectural features, will provide 80% of the annual heating for the north building and 75% of the heating of the south building. For both buildings 60% of the cooling will be accomplished with these systems.

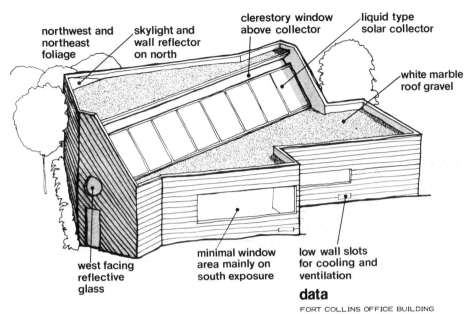

northwest and northeast foliage

skylight and wall reflector on north

clerestory window above collector

liquid type solar collector

white marble roof gravel

west facing reflective glass

minimal window area mainly on south exposure

low wall slots for cooling and ventilation

data

FORT COLLINS OFFICE BUILDING

Floor Area	3000 square feet
Collector Area	600 square feet
Solar Collector	Liquid type
Solar Collector Supplier	Undetermined
Heat Storage	1500 gallon water tank
Completed	Plans in progress
Architect	Lawrence C. Atkinson
	Crowther/Solar Group

This three tenant, three thousand square foot office building is scheduled to be constructed in Fort Collins, Colorado. The heat from a 600 square foot liquid type flat plate solar collector will be used for space heating and for heating domestic water. An antifreeze type liquid will circulate through the collector and receive heat from the absorber surface. The heated liquid will be piped to a heat exchanger, where the heat will be transferred to water before the liquid will return to the collector, and the water will be routed to a 1500 gallon storage tank.

For space heating, hot water will be pumped from storage through coils in a forced air distribution system. Domestic water will be heated using a heat exchanger through which water from the storage tank will be pumped on one side and domestic water will circulate on the other. Electricity is the auxiliary energy source for these systems.

The building will have minimal north and west side glazing. The west windows will be constructed with reflective glass to limit summer afternoon heat gains, yet allow a view. Double insulated glass set in non-conductive frames will be used exclusively. A double door airlock entry will provide protection against infiltration. Heat transmission through walls and roof will be minimized by insulation and a plio-film vapor barrier. Natural illumination will be provided through regular windows, clerestory windows above the solar collector, and a skylight with wall mounted reflector on the north side. Other optimized energy conservation features will be incorporated into the building design.

The solar energy collection system in conjunction with energy conservation is expected to provide 90% of annual space heating requirements and 95% of annual domestic hot water. Over 60% of the cooling is anticipated to be provided by natural ventilation.

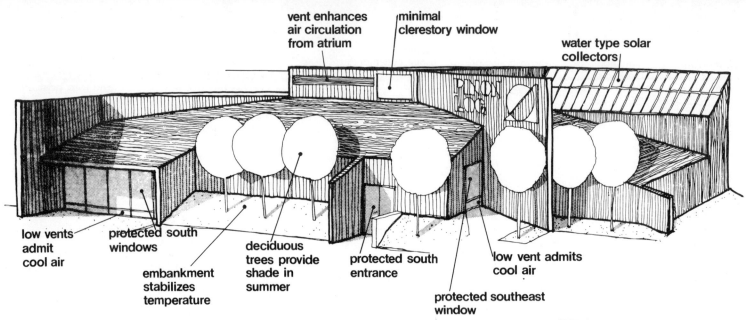

vent enhances
air circulation
from atrium

minimal
clerestory window

water type solar
collectors

low vents
admit
cool air

protected south
windows

embankment
stabilizes
temperature

deciduous
trees provide
shade in
summer

protected south
entrance

protected southeast
window

low vent admits
cool air

shopping center

Unique uses of solar and natural climatic energies in
conjunction with energy conservation will be in this
shopping center located near Pinery, Colorado. The
first phase has 33,000 square feet of business estab-
lishments plus another 12,000 square feet of covered
mall area. The buildings are designed for the site so
that a north embankment would buffer them from
winter winds and allow ample room for a parking
lot to the south. With this arrangement, primary
entrances to the mall are on the south or east sides
of the building where they are easily accessible from
the parking lot and would be sheltered from prevailing winds, minimizing heat gains and losses.

PINERY SHOPPING CENTER

Floor Area	45,000 square feet
Collector Area	8,000 square feet
Solar Collector	Liquid type
Solar Collector Supplier	Undetermined
Heat Storage	48,000 gallon mall pond
Completed	In design and planning stages
Architects	Richard L. Crowther AIA
	Lawrence C. Atkinson
	Crowther/Solar Group

The north side of the building has minimal fenestration; the doors here are for emergencies only. Entrances from all shops open onto the mall only. None open directly to the outside, providing a double or triple airlock system, which greatly decreases heat exchange by infiltration.

The successful operation of the heating and cooling systems of all the shops will depend upon a large decorative water pool located in the mall. A network of heat pumps will connect the shops thermally to the pool. The pool in turn is part of a system of water type solar collectors and cooling towers.

In the summer, the shops will be cooled by the heat pumps transporting excess heat from them to the pool in the mall. When the pool reaches a certain temperature, the water will be pumped to roof cooling towers, where the heat is given off to the air by evaporation, then routed back to the pond. The pond water must be relatively free from dissolved minerals so that plumbing is not corroded.

A day care center will be situated on the north side of the shopping center, which has an outside playground. The north wall of the playground will be covered with a specular reflective surface. The playground level will be raised slightly above the inside floor level, insuring that winter sunshine will clear the roof of the building, strike the mirror-like wall, and be directed into the playground. In this way, the northside area will receive direct winter warming. The height of the wall is calculated not to reflect sun directly into the eyes of children at play.

In the winter, the solar collectors will heat an antifreeze solution which will be piped to heat exchangers in the pool, adding heat to it. Water-to-air heat pumps will extract heat from the pool to warm the shops. If a shop with a southern exposure gets too hot, as a result of direct solar gains, mechanical equipment, or occupancy, the heat pump will remove the excess heat and deposit it in the pond, where it is available to heat another shop. The pool and its water fountain will provide the mall with heat, humidification, and beneficial negative ionization.

Southern windows will be sheltered with overhangs to block out summer sun, yet allow winter sun to enter. Clerestory windows will provide natural lighting to the mall with minimal heat loss and gain.

A second phase of the shopping center is designed to extend construction directly west of the first phase. It will have its own pool, solar collectors, and cooling towers with operating systems similar to the first phase.

Optimized energy conservation plus solar and climatic energies is expected to provide 80% of the annual heating and 60% of the annual cooling requirements of this shopping center.

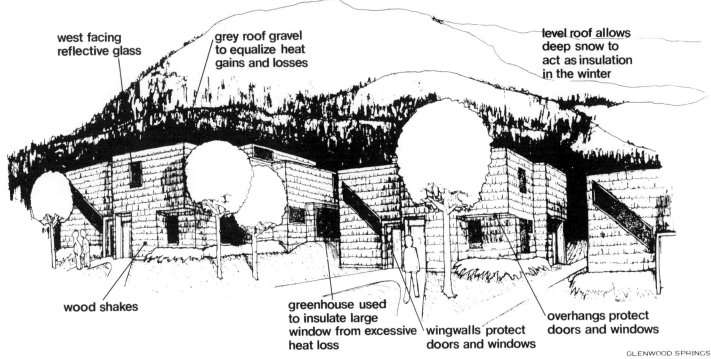

west facing
reflective glass

grey roof gravel
to equalize heat
gains and losses

level roof allows
deep snow to
act as insulation
in the winter

wood shakes

greenhouse used
to insulate large
window from excessive
heat loss

wingwalls protect
doors and windows

overhangs protect
doors and windows

GLENWOOD SPRINGS FOURPLEXES

Floor Area	1000 square feet per unit
Heat Source	Geothermal
Completed	Plans in progress
Architects	Lawrence C. Atkinson
	Crowther/Solar Group

Twenty-eight dwellings for the mountain community of Glenwood Springs, Colorado, have been designed with energy conservation features which are of particular value in this 9600 degree-day climate. Each individual dwelling is designed with approximately 1000 square feet of floor area, with at least one unit within each group of four having a different floor plan than the others.

When these buildings are completed in late 1976, resistance to heat loss will be a primary concern, and will be minimized by details of site planning, building design, and construction. Among the planned features of the site are multi-units under one building envelope, and clustering of the buildings for energy conservation. The thermal resistance values of all building envelope components will be maximized, within physical limitations imposed by the thickness of structural components, with walls insulated to R-30 (U-value of .033), roofs to R-54 (U-value of .018), and structural floors to R-44 (U-value of .023).

Although cloud cover is infrequent in Glenwood Springs, its location in a narrow valley limits sunlight to the point that active solar heat collection is uneconomic. Local geothermal energy potential is well-known, and its application to space heating, domestic water heating, and other uses is economically warranted on projects of this size.

Highly corrosive, hot water from the underground springs is planned for the heat source. Because of its corrosiveness, the water would deteriorate piping in a short length of time if pumped through it. For this reason, the hot spring water is brought to a heat exchanger at ground level where the heat is transferred to non-corrosive water in a closed-loop system. For space heating the hot non-corrosive water is pumped through well insulated pipes to zoned baseboard convection units in all the buildings. Domestic water will be heated in each building using heat exchangers through which the domestic water flows on one side, and the hot non-corrosive water on the other. Heated domestic water will be stored in insulated tanks, providing an abundant supply with instant availability.

Features which optimize energy conservation in these buildings include recesses, overhangs, and wingwalls to protect doors and windows from wind and regulate the entry of sunlight. Wood shake shingles covering exterior walls provide a low maintenance exterior which contributes to the minimization of heat loss and gain by controlling a still layer of air at the wall surface. Level roofs are designed to accumulate snow which will provide additional winter roof insulation. Grey roof gravel was selected to optimize a balance between year-round heat gains and losses.

Ventilation with the intake of outdoor air will be used for summer cooling. This will be provided through low-level wall vents, and the effect will be accentuated by low-level foliage around the building perimeters.

Geothermal energy in conjunction with optimized energy conservation is planned to supply 100% of the heating and hot water for these buildings. Since goethermal is a reliable source of energy, no auxiliary heating systems are needed. Cooling will be provided by natural ventilation with no back-up cooling system planned.

notes

Economics

Numerous solar, climatic energy systems, and architectural concepts presented in this book are immediately practical, technically sound, readily available, and cost effective, when measured against the unavailability of natural gas as a fuel, limitations on amounts of available electrical power, and the projected escalation in price or the existing high cost of conventional fuels. Other systems and concepts presented in this book lack sufficient engineering development, adequate performance data, and reasonable availability to be economically practical within a number of years.

For most people, the decision of whether or not to invest in energy optimized architecture, employ active or passive solar collection, or take advantage of other climatic energies is made purely from economic considerations. Unfortunately, the morality of both energy conservation and the long-term allocation of resources is seldom a determining factor in these decisions.

Historically, energy and resources have been viewed as abundant and therefore expendable commodities by the American consumer and designer. It was once more economic to install energy consuming mechanical equipment to create artificial indoor environments than to design the building and site to respond to local climatic energies for the creation of comfortable indoor environments.

Recent shortages have brought about the awareness that energy and resources are finite in their supplies and reserves. Short supplies coupled with high demand have caused prices to escalate, allocating energy and resources only to those who are able to pay the price. With each incremental price increase a proportionate number of people are ejected from the ranks of the "have" to those of the "have not". This economic selection process has encouraged increasing numbers of people to investigate alternative energy systems and regard energy conservation as a top priority.

Within the guidelines established by economics, energy and resource conservation are becoming increasingly attractive as the price of fossil fuels increases. Economic analysis tools, such as life-cycle costing, are replacing older initial cost concepts.

The economics of life-cycle costs are based on the life expectancy of a building and the total expenditures that will be incurred during that period for both initial or capital costs, plus operating and maintenance expenses. The first category includes all the front end costs of a project, everything from land and construction costs to architectural, engineering, and legal fees. The second category can include, in addition to annual operational, maintenance and repair expenses, such items as real estate taxes, insurance premiums and depreciation. An ever-increasing proportion of operational expenses is for the use of fuel.

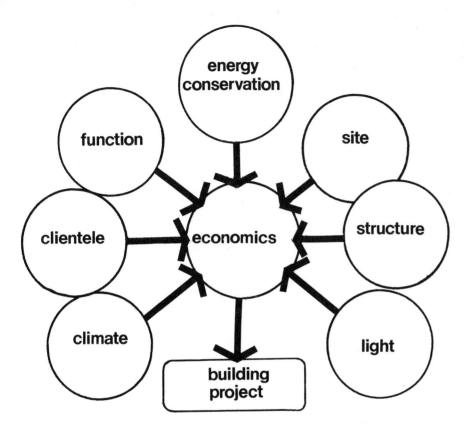

**holistic building
design process**

Because of the growing attrition rate for conventional energy sources, their cost is most apt to escalate by exponential rather than arithmetical increments as the year 2000 approaches, making life-cycle costing especially attractive to owners concerned with reducing the long term overall cost of owning a building project or home. These escalating fuel costs may ultimately prompt prudent tenants to investigate the heating, air-conditioning, and lighting expenses associated with their leased spaces, forcing even speculative commercial builders to consider operational and maintenance costs.

Most systems providing for optimized energy conservation and the use of climatic energies are capital intensive; that is, they have high front end or initial costs. Once they are operating they are anti-inflationary, for the cost of free energy never varies! As the long range costs of alternative energy sources are exceeded by the cost of each conventional fuel, life-cycle cost concepts can be applied with increasing success. The net result will be long term savings of money and a reduction in the overall use of resources.

Every location on earth is endowed with free climatic energy. The basic human requirement is how to either shelter from its extremes, directly use natural energy, or transform energy to moderate the local climate. The planning process for designing for unique microclimates and individual human needs and responses involves seven categories which revolve around the ultimate arbiter, economics. The inputs to functional design include clientele, climate, energy conservation, function, light, site, and structure. Each is dependent upon the others and is part of the holistic design process.

The use of natural energy sources is determined by their performance, practical installation, economic availability, and long-range suitability. The selective options are so diverse, interdependent, and bear such different socio-economic, environmental impact price tags that a systematic approach, design experience, technical knowledge, and frequently computer assistance are necessary to defined evaluation.

However, common sense will serve to keep one on the path toward energy efficient building with economic natural climatic utilization.

To realize a savings through optimized building design and site planning, the most cost effective priority order to a reduction in energy demand and use is:

a. Optimized space use, thermal efficiency, and climatic buffering of the home or building in conjunction with site topography, landscaping, and external functions.

b. Direct passive and active mechanical use of the clean alternative natural energies of the sun, wind, water, and earth.

c. A timeframe dependent operational and maintenance program that optimizes the reduction of conventional energy use.

d. Control over human factors that are contrary to the conservation of energy.

Depending on climate and location, various building types and occupancies are more or less well suited for cost effective energy conservation and for the use of climatic energy. Each project, therefore, requires a totally separate analysis to determine its suitability.

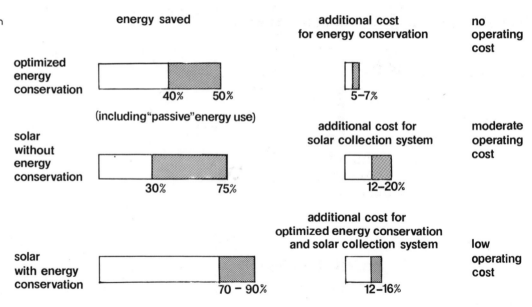

projected cost savings for commercial and institutional buildings*

energy saved

additional cost for energy conservation

no operating cost

optimized energy conservation — 40% 50%

(including "passive" energy use)

5-7%

solar without energy conservation — 30% 75%

additional cost for solar collection system

moderate operating cost

12-20%

additional cost for optimized energy conservation and solar collection system

low operating cost

solar with energy conservation — 70 - 90%

12-16%

*(space heating for Denver, Colorado, cost for 1976)

Every type of home and building occupancy has its own specific factor for total yearly energy demand. This demand is met by the use of energy. The more energy that can be derived from the natural climate, be it the sun, wind, temperature, humidity, precipitation, or earth, without the applied use of fuels or power, the lower will be the cost for energy use. The price of architecture or devices used to accumulate or directly use natural energies should be cost effective in terms of life-cycle analysis.

It is not cost effective to apply solar collection systems to poorly designed new homes and buildings. Solar collection systems simply have to be too large and expensive to serve homes and buildings that are thermally inefficient.

Solar collection economies in new construction depend more than 80% upon the cost effective optimization of energy conservation and passive use of natural climatic energy. The solar collector surface areas needed as a function of floor area to provide identical percentages of heating for a conventional building, one with minimal energy conservation (double glazing, extra insulation, weatherstripping, etc.), and one with optimized energy conservation are indicated on the next page for an average home and average office building. Optimized energy conservation significantly reduces the solar collector area required. At today's prices for solar collectors, an additional investment in optimized energy conservation is more than offset by the savings resulting from reduced required solar collector area. Surfaces such as mirrors, which are cheaper per square foot than solar collectors, if oriented to increase the incident radiation to a collector can help to decrease the required collector area, adding to the economy of the system.

It has proven possible through careful design, providing for the use of natural energy systems, to save 30 to 75% of the energy required for space heating in an optimized energy conserving residence when compared to a conventional residence, without adding more than 3 to 7% to the initial construction costs.

projected cost savings for single and multi-family residential buildings*

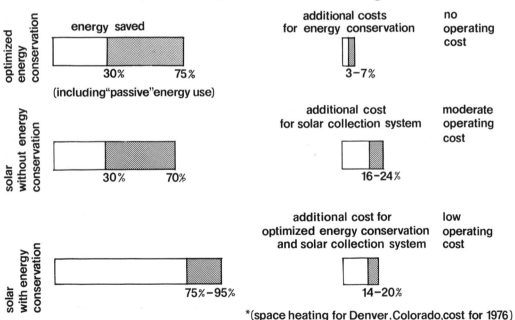

*(space heating for Denver, Colorado, cost for 1976)

When optimized energy conservation is combined with active solar collection, the size of the solar collector required may be economically scaled down. With optimized energy conservation, representing less than a 10% increase to the initial construction costs, the required solar collector area may be reduced by as much as 50% in size, with a savings of 80 to more than 90% of the space heating energy required by a typical conventional dwelling.

In new commercial and institutional projects it is possible to save 60% to 80% of space heating energy and 20 to 60% of the cooling energy, adding between 5% and 7% to initial construction costs for energy optimization and conservation design.

Because commercial buildings have large amounts of internally generated heat, they require less additional heat to maintain winter comfort than residences do. As a specific example, an office building with optimized energy conservation can require a solar collector area of only 10% of the building floor area. An optimized energy conserving home can require a solar collector area of 20% of the building floor area. In office buildings with optimized energy conservation and solar collection, space heating can be reduced by 70 to 90% and cooling by 40 to 60% when compared to conventional office buildings of similar size.

Homes and buildings vary greatly as to energy needs, depending on the occupancy and the specific character-istics of the microclimate. Multi-family structures and building mega-structures that have been efficiently planned offer a greater per capita opportunity for minimizing energy use than do single family homes or smaller individual commercial projects.

The cost of energy conservation through the use of earth, water, and air convection is most likely to be most economical, yet depends upon local climate, application, and geography. The direct conversion of wind to usable power is limited by geographic con-siderations, equipment costs, and the relative in-efficiency of battery storage.

As the economics of energy and resources become more definitive, building construction components that can be readily rearranged or reassembled on site and be used and reused for hundreds of years will emerge as elements that will greatly optimize the life-cycle of energy. Today, the projected life span of most new homes and buildings falls within the range of forty to seventy-five years. It should be 200 to 1000 years!

The world is presently in a period of dynamic social and economic revolution. The scarcity of energy is reshaping economic, social, and political institutions. Nature has provided inexhaustible supplies of solar and climatic energies which are free to those who choose to utilize them. The wisdom of this choice becomes more obvious with each incremental increase in the price of gasoline, coal, fuel oil, natural gas, and other energy and products derived from fossil fuels.

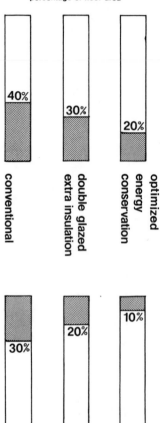

size of solar collector
(Denver)

average house
percentage of floor area

40% conventional

30% double glazed extra insulation

20% optimized energy conservation

30% conventional

20%

10%

average office
percentage of floor area

It is interesting to see where energy is consumed in our society. The following lists some end uses of energy.

END USES OF ENERGY

(consuming sector of the economy in parentheses)	Percent of Total
Transportation (fuel, excludes lubes, and greases)	24.9
Space heating (residential and commercial)	17.9
Process steam (industrial)	16.7
Direct heat (industrial)	11.5
Electric drive (industrial)	7.9
Raw materials and feedstocks (commercial, industrial, and transportation)	5.5
Water heating (residential, commercial)	4.0
Air Conditioning (residential, commercial)	2.5
Refrigeration (residential, commercial)	2.2
Lighting (residential, commercial)	1.5
Cooking (residential, commercial)	1.3
Electrolytic processes (industrial)	1.2
Other	2.9
Total	100 percent

Energy Source	Energy Output	
1 ton bituminous coal	7674 kilowatt-hours	26.2×10^6 Btu
1 barrel crude oil	1640 kilowatt-hours	5.6×10^6 Btu
1 barrel residual oil	1842 kilowatt-hours	6.29×10^6 Btu
1 gallon gasoline	36 kilowatt-hours	125×10^3 Btu
1 gallon No. 2 fuel oil	40 kilowatt-hours	139×10^3 Btu
1 cubic foot natural gas	.301 kilowatt-hours	1031 Btu
1 kw-hr electricity	1.0 kilowatt-hours	3413 Btu
1 Ton TNT	11657 kilowatt-hours	39.8×10^6 Btu
1 megaton nuclear bomb	11,657,000,000 kilowatt-hours	39.8×10^{12} Btu
1 gram of matter completely converted to energy	24,985,354 kilowatt-hours	85.3×10^9 Btu

The following conversions are useful for design purposes, and will be especially helpful when the S.I. units of measurement are adopted in the USA.

energy

1 British Thermal Unit (Btu)	= 251.99 calories
	= 1055.06 joules
	= .00029287 kilowatt-hours
1 calorie	= .003968 Btu
1 foot-pound	= .324048 Calories
1 joule	= 1 watt sec
1 kilowatt-hour	= 3414.43 Btu

energy density

1 calorie/sq. cm.	= 3.68669 Btu/sq. ft.
1 Btu/square foot	≡ .271246 calories/sq. cm.
1 langley	= 1 calorie/sq. cm.

power density

1 cal./sq. cm./min.	= 221.2 Btu/sq. ft./hour
1 watt/sq. cm.	= 3172 Btu/sq. ft./hour

power

1 Btu/hour	= 4.2 calories/minute
	= .292875 watts
1 watt	= 1 joule/sec

flow rate

1 cubic foot/minute	= 471.947 cubic cm/second
1 liter/minute	= .0353 cubic feet/minute
	= .2642 gallons/minute

mass/weight

1 pound	= 16 ounces
	= .45359 kilograms
1 ton	= 907 kilograms
1 kilogram	= 2.2046 pounds
1 metric ton	= 1000 kilograms
	= 2204.6 pounds

velocity

1 foot/minute	= .508 centimeter/second
1 mile/hour	= 1.6093 kilometer/hour
1 kilometer/hour	= .621 mph

temperature

$$^{\circ}C = 5/9\ (^{\circ}F - 32)$$

$$^{\circ}F = 9/5\ (^{\circ}C) + 32$$

Metric Prefix	Common Usage	Scientific Notation
nano	1 billionth	10^{-9}
micro	1 millionth	10^{-6}
milli	1 thousandth	10^{-3}
centi	1 hundredth	10^{-2}
kilo	1 thousand	10^{3}
mega	1 million	10^{6}

length

1 mile	= 5280 feet
	= 1.6093 kilometers
	= 1760 yards
1 kilometer	= .621 miles
	= 1000 meters
1 yard	= .9144 meters
1 meter	= 39.37 inches
	= 3.28 feet
1 centimeter	= .3937 inch
1 inch	= 2.54 centimeters
1 foot	= .3048 meter
1 angstrom	= 1×10^{-8} centimeters

area

1 square mile	= 640 acreas
	= 2.59 square kilometers.
1 square yard	= .836 square meters
1 square foot	= .0929 square meters
1 square inch	= 6.4516 square centimeters
1 square centimeter	= .155 square inches
1 square meter	= 10.7639 square feet
	= 1.196 square yards
1 square kilometer	= .3861 square miles
1 acre	= 43,560 square feet
	= 4047 square meters

volume·liquid

1 gallon	= 4 quarts
	= 3.7854 liters
	= 231 cubic inches
1 quart	= .9463 liters
1 liter	= 1000 cubic centimeters
	= 1.0567 quarts
	= .2642 gallons

volume·dry

1 cubic foot	= 28.317 liters
1 cubic yard	= .7645 cubic meter
1 cubic inch	= 16.387 cubic centimeters
1 cubic centimeter	= .06102 cubic inches
1 cubic meter	= 35.3145 cubic feet
	= 1000 liters
	= 1.308 cubic yards

The following maps present the mean daily insolation on a horizontal surface for the months of June and December and year-round. The unit of measurement is langleys, where 1 langley/day = 1 calorie/sq. cm./day (3.69 Btu/sq. ft./day).

The source of these maps is the U. S. Department of Commerce, Weather Bureau. Copies are sold by The Superintendent of Documents, U. S. Government Printing Office, Washington, D.C. 20402.

Mean Daily Solar Radiation in June (langleys)

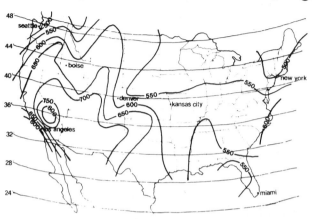

Mean Daily Solar Radiation in December (langleys)

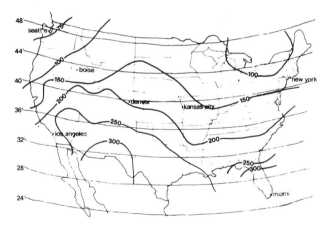

Mean Daily Solar Radiation Annual (langleys)

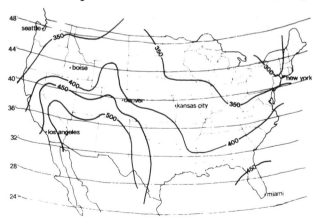

The following maps present the mean monthly hours of sunshine for the months of June and December and year-round.

The source of these maps is the U. S. Department of Commerce, Weather Bureau. Copies are sold by The Superintendent of Documents, U. S. Government Printing Office, Washington, D.C. 20402.

Mean Monthly Hours of Sunshine Annually

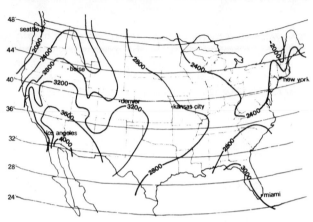

Mean Monthly Hours of Sunshine in June

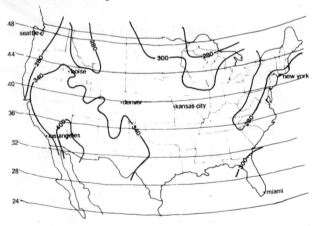

Mean Monthly Hours of Sunshine in December

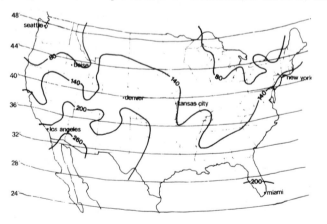

This map presents the average annual wind power (watts/square meter) which is available in the USA.

Average Annual Wind Power (watts/sq.meter)

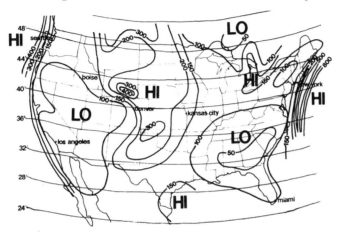

The following maps, based on mean daily insolation, present the relative suitability for winter and year-round solar energy collection.

Relative Suitability for Solar Energy Collection (winter)

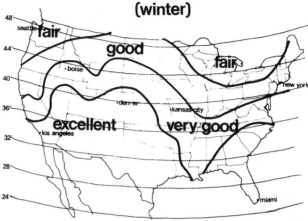

Relative Suitability for Solar Energy Collection (year-round)

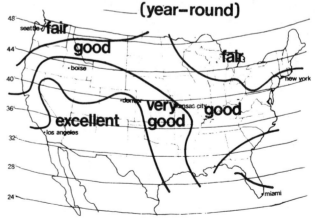

Used with the permission of Revere Copper and Brass, Inc.

The following three charts present the average number of Btu's per square foot per day striking surfaces tilted at various angles in three geographic locations. The locations are Denver, Colorado; Nashville, Tennessee; and Ithaca, New York. Both climatic conditions and latitude of the locations were used to prepare the charts. These three cities were chosen for presentation because they represent divergent locations and also have unique quantities of available radiation. Data is available which can be used to prepare similar charts for most metropolitan areas in the USA.

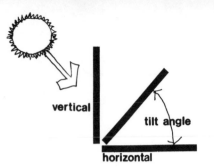

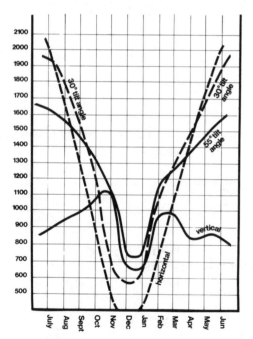

Btu's / sq. ft. / day – Ithaca, New York

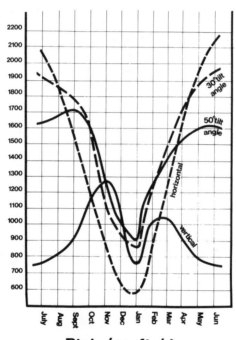

Btu's / sq. ft. / day
Nashville, Tennessee

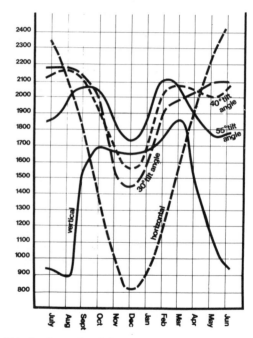

Btu's / sq. ft. / day – Denver, Colorado

The following are lists of the sun's location in the sky relative to various northern latitudes. The altitude is measured in degrees above horizontal and the azimuth is measured in degrees east or west of true south. The time is referenced to the sun with noon occurring when the azimuth angle is zero degrees.

This information is valuable for calculating the recess depths of windows or for sizing wingwalls and overhangs where sun control to the interior is desired. It can be used for making most determinations relative to building components and the sun's position.

azimuth

N

altitude

Latitude N 30°

AM	PM		JAN 21	FEB 21	MAR 21	APR 21	MAY 21	JUN 21	JUL 21	AUG 21	SEP 21	OCT 21	NOV 21	DEC 21
5	7	ALT												
		AZI												
6	6	ALT												
		AZI												
7	5	ALT	2	7	13	19	22	24	23	19	13	7		
		AZI	65	73	82	93	101	104	101	94	82	73		
8	4	ALT	14	19	26	32	35	37	36	32	26	20	14	11
		AZI	57	64	74	85	94	98	95	86	74	65	57	54
9	3	ALT	24	31	38	44	48	50	49	45	38	31	24	21
		AZI	47	54	63	76	87	92	88	77	63	54	47	44
10	2	ALT	32	40	49	57	61	63	61	57	49	40	32	29
		AZI	34	40	49	63	77	83	78	64	49	40	34	32
11	1	ALT	38	47	57	67	73	75	74	67	57	47	38	35
		AZI	18	22	28	40	57	67	59	41	28	22	18	17
12		ALT	40	49	60	72	80	83	81	72	60	50	40	37
		AZI	0	0	0	0	0	0	0	0	0	0	0	0

Latitude N 26°

AM	PM		JAN 21	FEB 21	MAR 21	APR 21	MAY 21	JUN 21	JUL 21	AUG 21	SEP 21	OCT 21	NOV 21	DEC 21
5	7	ALT												
		AZI												
6	6	ALT												
		AZI												
7	5	ALT	4	8	13	18	22	23	22	19	13	9	4	2
		AZI	65	74	83	94	102	106	103	95	83	74	66	62
8	4	ALT	16	21	27	32	35	36	35	32	27	21	16	14
		AZI	58	66	76	88	94	101	98	88	76	66	58	55
9	3	ALT	27	33	39	45	48	49	49	46	39	33	27	24
		AZI	48	56	66	80	91	96	92	81	66	56	48	45
10	2	ALT	36	43	51	58	62	63	62	59	51	43	36	33
		AZI	35	42	53	69	84	91	85	70	53	42	35	33
11	1	ALT	42	50	60	70	75	76	75	70	60	51	42	38
		AZI	19	24	31	47	70	83	72	49	31	24	19	18
12		ALT	44	53	64	76	84	87	85	76	64	54	44	41
		AZI	0	0	0	0	0	0	0	0	0	0	0	0

Latitude N 32°

AM	PM		JAN 21	FEB 21	MAR 21	APR 21	MAY 21	JUN 21	JUL 21	AUG 21	SEP 21	OCT 21	NOV 21	DEC 21
5	7	ALT												
		AZI												
6	6	ALT				6	10	12	11	7				
		AZI				100	107	110	108	101				
7	5	ALT	1	7	13	19	23	24	23	19	13	7	2	
		AZI	65	73	82	92	100	103	101	93	82	73	65	
8	4	ALT	13	19	25	31	35	37	36	32	25	19	13	10
		AZI	57	64	73	84	93	97	94	85	73	64	57	54
9	3	ALT	23	29	37	44	48	50	48	44	37	30	23	20
		AZI	46	53	62	74	85	89	86	75	62	53	46	44
10	2	ALT	31	39	47	56	61	62	61	56	47	39	31	28
		AZI	33	39	48	60	73	80	74	61	48	39	33	31
11	1	ALT	36	45	55	65	72	74	72	66	55	45	36	33
		AZI	18	21	27	38	52	61	53	38	27	21	18	16
12		ALT	38	47	58	70	78	82	79	70	58	48	38	35
		AZI	0	0	0	0	0	0	0	0	0	0	0	0

Latitude N 34°

SOLAR TIME AM	PM		JAN 21	FEB 21	MAR 21	APR 21	MAY 21	JUN 21	JUL 21	AUG 21	SEP 21	OCT 21	NOV 21	DEC 21
5	7	ALT												
		AZI												
6	6	ALT												
		AZI												
7	5	ALT		6	12	19	23	25	23	19	12	6		
		AZI		73	81	92	99	103	100	92	81	73		
8	4	ALT	11	18	24	31	36	37	36	32	24	18	12	9
		AZI	56	63	72	83	91	95	92	83	72	63	56	54
9	3	ALT	21	28	36	43	48	49	48	44	36	28	21	18
		AZI	45	52	61	72	82	87	83	73	61	52	46	43
10	2	ALT	29	37	46	55	60	62	60	55	46	37	29	26
		AZI	32	38	46	58	70	76	71	59	46	38	33	31
11	1	ALT	34	43	53	64	71	73	71	64	53	43	34	31
		AZI	17	20	26	35	47	55	49	36	26	20	17	16
12		ALT	36	45	56	68	76	79	77	68	56	46	36	33
		AZI	0	0	0	0	0	0	0	0	0	0	0	0

Latitude N 38°

SOLAR TIME AM	PM		JAN 21	FEB 21	MAR 21	APR 21	MAY 21	JUN 21	JUL 21	AUG 21	SEP 21	OCT 21	NOV 21	DEC 21
5	7	ALT												
		AZI												
6	6	ALT				7	12	14	13	8				
		AZI				99	106	109	106	100				
7	5	ALT		5	12	19	24	26	24	19	12	5		
		AZI		72	81	90	98	101	98	91	81	72		
8	4	ALT	9	16	23	31	36	37	36	31	23	16	9	7
		AZI	56	62	70	80	89	92	89	81	70	62	56	53
9	3	ALT	18	26	34	42	47	49	48	42	34	26	18	15
		AZI	44	50	58	69	78	82	79	70	58	51	45	42
10	2	ALT	26	34	43	52	58	61	59	53	43	34	26	22
		AZI	31	36	43	53	64	69	65	54	43	36	31	30
11	1	ALT	30	39	50	60	68	71	68	61	50	40	31	27
		AZI	16	19	24	31	40	46	41	31	24	19	16	15
12		ALT	32	41	52	64	72	75	73	64	52	42	32	29
		AZI	0	0	0	0	0	0	0	0	0	0	0	0

Latitude N 36°

SOLAR TIME AM	PM		JAN 21	FEB 21	MAR 21	APR 21	MAY 21	JUN 21	JUL 21	AUG 21	SEP 21	OCT 21	NOV 21	DEC 21
5	7	ALT						2						
		AZI						117						
6	6	ALT				7	12	14	12	7				
		AZI				99	106	109	106	99				
7	5	ALT		6	12	19	24	25	24	19	12	6		
		AZI		72	81	91	98	101	98	91	81	72		
8	4	ALT	10	17	24	31	36	38	36	31	24	17	10	8
		AZI	56	62	71	81	90	93	90	81	71	62	56	53
9	3	ALT	20	27	35	42	48	49	48	42	35	27	20	17
		AZI	44	51	59	70	80	84	80	70	59	51	44	42
10	2	ALT	27	35	45	54	60	62	60	54	45	35	27	24
		AZI	32	36	44	55	67	72	67	55	44	36	32	30
11	1	ALT	32	41	52	62	69	72	69	62	52	41	32	28
		AZI	17	20	24	33	43	49	43	33	24	20	17	15
12		ALT	34	43	54	66	74	77	74	66	54	43	34	30
		AZI	0	0	0	0	0	0	0	0	0	0	0	0

Latitude N 40°

SOLAR TIME AM	PM		JAN 21	FEB 21	MAR 21	APR 21	MAY 21	JUN 21	JUL 21	AUG 21	SEP 21	OCT 21	NOV 21	DEC 21
5	7	ALT				2	4	2						
		AZI				115	117	115						
6	6	ALT				7	13	15	13	8				
		AZI				99	106	108	106	100				
7	5	ALT		4	11	19	24	26	24	19	11	5		
		AZI		72	80	90	97	100	97	90	80	72		
8	4	ALT	8	15	23	30	35	37	36	31	23	15	8	6
		AZI	55	62	70	79	87	91	88	80	70	62	55	53
9	3	ALT	17	24	33	41	47	49	47	42	33	25	17	14
		AZI	44	50	57	67	76	80	77	68	57	50	44	42
10	2	ALT	24	32	42	51	58	60	58	52	42	32	24	21
		AZI	31	35	42	51	61	66	62	52	42	36	31	29
11	1	ALT	28	37	48	59	66	69	67	59	48	38	29	25
		AZI	16	19	23	29	37	42	78	30	23	19	16	15
12		ALT	30	39	50	62	70	74	71	62	50	40	30	27
		AZI	0	0	0	0	0	0	0	0	0	0	0	0

Latitude N 42°

AM	PM		JAN 21	FEB 21	MAR 21	APR 21	MAY 21	JUN 21	JUL 21	AUG 21	SEP 21	OCT 21	NOV 21	DEC 21
5	7	ALT												
		AZI												
6	6	ALT				8	13	15	14	8				
		AZI				99	105	108	106	99	90			
7	5	ALT			4	11	19	24	26	25	19	11	4	
		AZI			72	80	89	96	99	96	89	80	72	
8	4	ALT	7	14	22	30	35	37	36	30	22	14	7	4
		AZI	55	61	69	78	86	89	86	79	69	61	55	53
9	3	ALT	15	23	32	40	46	48	47	41	32	23	16	12
		AZI	44	49	56	66	74	78	75	66	56	49	44	42
10	2	ALT	22	30	40	50	56	59	57	50	40	31	22	19
		AZI	30	35	41	50	58	63	59	50	41	35	31	29
11	1	ALT	26	35	46	57	65	68	65	58	46	36	27	23
		AZI	16	18	22	28	35	39	35	28	22	18	16	15
12		ALT	28	37	48	60	68	71	69	60	48	38	28	25
		AZI	0	0	0	0	0	0	0	0	0	0	0	0

Latitude N 46°

AM	PM		JAN 21	FEB 21	MAR 21	APR 21	MAY 21	JUN 21	JUL 21	AUG 21	SEP 21	OCT 21	NOV 21	DEC 21
5	7	ALT												
		AZI												
6	6	ALT				8	14	17	15	9				
		AZI				98	104	107	105	99				
7	5	ALT		2	10	19	25	27	25	19	10	3		
		AZI		72	79	87	94	97	94	88	79	72		
8	4	ALT	5	12	20	29	35	37	35	30	20	12	5	2
		AZI	55	60	67	76	83	86	84	76	67	61	55	53
9	3	ALT	12	20	29	39	45	47	45	39	29	21	13	9
		AZI	43	48	54	63	70	74	71	63	54	48	43	41
10	2	ALT	19	27	37	47	54	57	55	48	37	27	19	15
		AZI	30	33	39	46	53	57	54	47	39	34	30	28
11	1	ALT	23	32	42	53	61	64	62	54	42	32	23	19
		AZI	15	17	20	25	30	33	31	25	20	17	15	15
12		ALT	24	33	44	56	64	67	65	56	44	34	24	21
		AZI	0	0	0	0	0	0	0	0	0	0	0	0

Latitude N 44°

AM	PM		JAN 21	FEB 21	MAR 21	APR 21	MAY 21	JUN 21	JUL 21	AUG 21	SEP 21	OCT 21	NOV 21	DEC 21
5	7	ALT					4	6	4					
		AZI					115	117	115					
6	6	ALT				8	14	16	14	8				
		AZI				98	105	107	105	98				
7	5	ALT			3	11	19	24	27	24	19	11	3	
		AZI			72	80	88	95	98	95	88	80	72	
8	4	ALT	6	13	22	30	35	37	35	30	22	13	6	3
		AZI	55	61	68	77	85	88	85	77	68	61	55	53
9	3	ALT	13	22	31	40	46	48	46	40	31	22	13	11
		AZI	43	48	55	64	72	76	72	64	55	48	43	41
10	2	ALT	20	28	38	49	55	58	55	49	38	28	20	17
		AZI	30	34	40	48	56	60	56	48	40	34	30	29
11	1	ALT	24	33	44	55	63	66	63	55	44	33	24	21
		AZI	15	18	21	27	33	36	33	27	21	18	15	15
12		ALT	26	35	46	58	66	69	66	58	46	35	26	22
		AZI	0	0	0	0	0	0	0	0	0	0	0	0

Latitude N 48°

AM	PM		JAN 21	FEB 21	MAR 21	APR 21	MAY 21	JUN 21	JUL 21	AUG 21	SEP 21	OCT 21	NOV 21	DEC 21
5	7	ALT					5	8	6					
		AZI					114	117	115					
6	6	ALT				9	15	17	15	9				
		AZI				98	104	106	104	98				
7	5	ALT		2	10	19	25	27	25	19	10	2		
		AZI		72	79	87	93	96	94	87	79	72		
8	4	ALT	4	11	20	29	35	37	35	29	20	11	4	
		AZI	55	60	67	75	82	85	82	75	67	60	55	
9	3	ALT	11	19	28	38	44	47	45	38	28	19	11	8
		AZI	43	47	53	61	68	72	69	62	53	47	43	41
10	2	ALT	17	26	35	46	53	56	54	46	35	26	17	14
		AZI	29	33	38	45	51	55	52	45	38	33	30	28
11	1	ALT	21	30	40	52	60	63	60	52	40	30	21	17
		AZI	15	17	20	24	29	31	29	24	20	17	15	14
12		ALT	22	31	42	54	62	65	63	54	42	32	22	19
		AZI	0	0	0	0	0	0	0	0	0	0	0	0

The following is a list of some of the most common household appliances and the average wattage of each. This list can be used to estimate the amount of energy and ultimately the amount of money that is spent each year for their operation. The procedure is to first estimate the number of hours per year each individual appliance is operated, use the average wattage from the list to determine the annual number of kilowatt-hours (Kwh) of electricity consumed by the appliance, then with the cost per Kwh of electricity figure the yearly cost of operation.

For example, estimate that a broiler is used for 3 hours per week, then the average yearly usage is: 3 x 52 = 156 hours. During operation, the broiler requires 1430 watts. For the year the total consumption is 1430 x 156 = 223,080 watt-hours. Dividing by 1000, the consumption is: 223,080/1000 = 223 Kwh.

If the price for electricity is 4¢/Kwh, then the annual cost to operate the broiler is $8.92. Summing the costs to run individual appliances will yield the total annual cost for operating all household appliances.

Appliance	Average Wattage
Air cleaner	50
Air conditioner	1566
Blanket, electric	180
Blender	350
Bottle sterilizer	500
Bottle warmer	500
Broiler	1430
Carving knife	90
Clock	2
Clothes dryer, electric heat	4856
Clothes dryer, gas heat	325
Coffee pot	890
Corn popper	460-550
Curling iron	10-20
Deep fryer	1440
Dehumidifier	300-500
Dishwasher	1200
Disposal	375
Drill, electric	250
Egg cooker	510
Fan, attic	370
Fan, circulating	88
Fan, roll away	171
Fan, window	200
Food warming tray	350
Foot warmer	75
Floor polisher	300

Appliance	Average Wattage
Freezer, 15 cubic feet	340
Freezer, 15 cubic feet, frost free	440
Frying pan	1190
Germicidal lamp	20
Griddle	700
Grill	1000
Hair dryer	400
Heat lamp, infrared	250
Heater, portable	1300
Heating pad	70
Hot plate	1250
Humidifier	177
Iron, hand	1100
Knife sharpener	125
Mixer, food	120
Movie projector	500
Oven, Microwave	1450
Radio, table	60
Radio, record player	100
Range, with oven	12200
Refrigerator/freezer 14 cu. ft.	320
Refrigerator/freezer frost free, 14 cu. ft.	610
Roaster	1300
Rotisserie	1400
Sandwich grille	1160
Sewing machine	75

Appliance	Average Wattage
Shaver	12
Stereo system	200
Sun lamp	300
Toaster	1150
Toothbrush, electric	7
Trash compactor	400
TV - black/white, tube type	160
TV - black/white, solid state	55
TV - color, tube type	300
TV - color, solid state	200
Typewriter, electric	40
Vacuum cleaner	600
Vaporizer	300
Vibrator	40
Waffle iron	1100
Washing machine, automatic	500
Washing machine, non-automatic	280
Waste disposer	440
Water heater	2470
Water heater, quick recovery	4474
Water pump, electric	450

The annotated bibliography which follows contains the references used in the preparation of this book, plus other publications which can be of value for exploring concepts which are beyond the scope of those presented here. The bibliography is divided into major categories to facilitate access to publications presenting specific topics, with comments on each entry provided to expedite the selection process.

energy-general
alternative energy sources

THE BIOSPHERE. Scientific American. W.H. Freeman and Company, 1970. 134 pages.
    A collection of 11 articles on man and his biosphere.

COPING WITH THE ENERGY CRISIS. Consumer Federation of American OEO, 1974. 81 pages.
    An Office of Economic Opportunity report on efforts of agencies and community groups to adjust to the energy situation.

THE COSTS OF SPRAWL. Detailed Cost Analysis. Real Estate Research Corporation, CEQ, HUD, EPA, 1974. 278 pages.
    A report on the energy, environmental, and human effects of low density sprawl.

EDISON EXPERIMENTS. Robert F. Schultz. Thomas Alva Edison Foundation, 1969. 30 pages.

ENVIRONMENTAL EXPERIMENTS...FROM EDISON. Robert F. Schultz. Thomas Alva Edison Foundation, 1973. 32 pages.
    Two small booklets with examples and experiments to illustrate environmental and energy concerns.

ENERGY. AIA Energy Notebook. American Institute of Architects, 1975.
    On going notebook with reference to books, articles, and regulations involving energy and its use.

ENERGY / LIFE SCIENCE LIBRARY. Mitchell Wilson. Time-Life Books, 1968. 198 pages.
    General energy book covering principles and application of various energy sources.

ENERGY: USE, CONSERVATION AND SUPPLY. Philip H. Abelson. American Association for the Advancement of Science, 1974. 155 pages.
    Twenty four articles on new and developing energy approaches.

ENERGY DEVELOPMENT IN THE ROCKY MOUNTAIN REGION: GOALS AND CONCERNS. Federation of Rocky Mountain States, Inc., 1975. 134 pages.
    Summary of studies in a 5 state region on energy utilization and implications of future resource development.

ENERGY EARTH AND EVERYONE. A Global Strategy for Spaceship Earth. Medard Gabel. Straight Arrow Books, San Francisco, 1975. 160 pages.
    A book developed by the World Game Workshop in Connecticut, including a foreword by R. Buckminster Fuller. Presents an analysis of worldwide existing and alternative energy sources.

ENERGY FOR SURVIVAL. The Alternative to Extinction. Wilson Clark. Anchor Press/Doubleday, 1975. 652 pages.
    This book not only examines the advantages and disadvantages of fuels currently in use, but also details numerous natural energy principles. The energy habits and policy of the American society are examined.

ENERGY AND POWER. Scientific American. W. H. Freeman and Company, 1971. 144 pages.
    A collection of 11 articles on the general use of energy and its effect on various cultures.

ENERGY PRIMER. Solar, Water, Wind and Biofuels. Portola Institute. Fricke-Parks Press, 1974. 200 pages.
    A collection of articles on solar, wind, water, architecture, and integrated energy systems with a "Whole Earth" style of catalog for energy related hardware and information.

ENERGY BOOK #1. Natural Sources and Backyard Applications. Edited by John Previs. 112 pages.
    Collection of articles and information on wind, solar, and biosystems energy generation.

ENERGY AND THE FUTURE. Allen L. Hammond. William D. Metz and Thomas H. Maugh II. American Association for the Advancement of Science, 1973. 184 pages.
    Analysis of various energy sources and the implications of their use.

FUTURE ENERGY OUTLOOK. Quarterly of the Colorado School of Mines, 1973. 206 pages.
    1972 Proceedings of Mineral Economics Symposium and the 1968 Proceedings of the Fuels Symposium of the American Association of Petroleum Geologists.

THE HOUSEHOLD ENERGY GAME. Thomas W. Smith and John Jenkins. University of Wisconsin – Madison, Wisconsin, 1974. 20 pages.
    A game to determine personal energy consumption in the home and by the family car.

THE NATION'S ENERGY FUTURE. Dr. Dixy Lee Ray, United States Atomic Energy Commission, 1973. 171 pages.
    A report on existing energy resources and projected uses.

NATURAL ENERGY WORKBOOK. Peter Clark. Visual Purple, 1974. 98 pages.
    Presents basic concepts related to solar, wind, and other alternative energy source systems.

MAN AND THE ECOSPHERE. Scientific American. W. H. Freeman and Company, 1971. 307 pages.
    A collection of 27 articles on man's ecosphere and the environmental effects of his actions.

PRINCIPLES OF HEAT TRANSFER. Frank Kreith. Intext Educational Publishers, 1973. 656 pages.
    A textbook presenting engineering and technical aspects of thermal conduction, radiation, and convection.

SURVIVAL 2001, Scenario from the future. Henry Voegeli and John J. Tarrant. Van Nostrand Reinhold Company, 1975. 115 pages.
    A book which presents the alternative energy sources; wind, solar, water, wave, and ocean power, using a futuristic standpoint.

energy conservation

ENERGY CONSERVATION IN ARCHITECTURE. Part I: ADAPTING DESIGN TO CLIMATE. Donald Watson in Connecticut Architect, March/April 1974.

ENERGY CONSERVATION IN ARCHITECTURE. Part II: ALTERNATIVE ENERGY SOURCES. Donald Watson and Everett Barber, Jr. in Connecticut Architect, May/June 1974. 13 pages.
    Lists measures to be taken for energy conservation in the design of buildings. Briefly covers alternative energy sources.

ENERGY CONSERVATION IN BUILDING DESIGN. American Institute of Architects, AIA Research Corporation, 1974. 156 pages.
    Included in this book are design methods for maximizing energy efficiency in building design. Also included are analyses of energy uses in existing buildings.

ENERGY CONSERVATION IN BUILDINGS; TECHNIQUES FOR ECONOMICAL DESIGN. C. W. Griffin, Jr. The Construction Specifications Institute, Inc., 1974. 183 pages.
    Presents principles of mechanical systems and building design necessary for energy conserving structures.

ENERGY CONSERVATION THROUGH DESIGN. Fred S. Dubin in Professional Engineer, page 21; October 1973.

ENERGY CONSERVATION DESIGN GUIDELINES FOR OFFICE BUILD-INGS. Dubin-Mindell-Bloome Associates, AIA Research Corporation, Heery and Heery, Architects. GSA/PBS, 1974.

ENERGY CONSERVATION STUDIES FOR NEW AND EXISTING BUILDINGS. Fred S. Dubin, Dubin-Mindell-Bloome Associates, P.C., 1975.
Articles and books on energy conservation methods and implications. Most of the energy consuming aspects of buildings are analyzed.

ENERGY CONSERVATION AND THE BUILDING SHELL. Building Systems Information Clearinghouse, 1974. 28 pages.
Gives examples of energy conserving measures which can be taken on new and existing buildings plus an analysis of their long-term economic effects.

ENERGY CONSERVATION IN TRACT HOUSING. Clovis Heimsath Associates. AIA Research Corporation, 1974. 60 pages.
A book which examines the effect of incorporating energy conserving modifications into existing tract housing.

ENERGY CONSERVATION PROGRAM GUIDE FOR INDUSTRY AND COMMERCE. R. R. Gatts, R. G. Massey, and John C. Robertson. United States Department of Commerce, NBS Handbook #115, 1974. Approximately 200 pages.
Proposes new standards to American commerce and industry for decreasing their total energy consumption.

ENERGY, ENVIRONMENT AND BUILDING. Philip Steadman. Cambridge University Press, 1975. 287 pages.
This book includes articles on solar and wind energy, plus methods for energy conservation in buildings. Numerous examples and excellent descriptions of energy conserving buildings in the United States are included.

ENERGY AND FORM. An Ecological Approach to Urban Growth. Ralph Knowles. MIT Press, 1974.
A worthwhile book on the historical analysis of primitive architecture and design concepts. The study includes the development of forms which respond to natural and climatic stimuli.

EVALUATING LEGISLATION FOR ENERGY CONSERVATION IN BUILDING. American Institute of Architects, 1975.
Examines the architectural and energy implications of new legislative actions. Develops criteria for other legislative measures.

OTHER HOMES AND GARBAGE. Designs for Self-Sufficient Living. Leckie, Masters, Whitehouse and Young. Sierra Club Books, 1975.
Presents the basic concepts related to solar, wind, bioconversion, and other alternative energy systems.

TECHNICAL OPTIONS FOR ENERGY CONSERVATION IN BUILDING. United States Department of Commerce, NBS, 1973. 175 pages.
Investigates the energy conserving options for new and existing buildings. Several energy conservation features are also discussed.

architecture

DESIGN WITH CLIMATE. Bioclimatic Approach to Architectural Regionalism. Victor Olgyay. Princeton University Press, 1963. 190 pages.
The design of buildings to respond to the natural forces of climate and the sun. Examples of many different project sizes and applications.

EARTH INTEGRATED ARCHITECTURE. College of Architecture, Arizona State University, 1974.
A notebook covering the applications and implementation of below grade architectural principles.

A GUIDE TO SITE AND ENVIRONMENTAL PLANNING. Henry M. Rubenstein. John Wiley & Sons, Inc., 1969. 190 pages.
An analysis of landscaping, site planning, and environmental concerns.

A LANDSCAPE FOR HUMANS. Peter van Dresser. Biotechnic Press, 1972. 128 pages.
Examines the rationale behind regional and local planning concepts and the implications of more human oriented planning processes.

MASTER BUILDERS OF THE ANIMAL WORLD. David Hancocks. Harper & Row, 1973.
This book examines the design habits and dwellings developed by animals in response to their individual climates and needs.

MECHANICAL AND ELECTRICAL EQUIPMENT FOR BUILDINGS. Fifth Edition. William J. McGuinness and Benjamin Stein. John Wiley and Sons, 1974. 1011 pages.
A textbook covering the design, techniques, and construction of mechanical systems for buildings.

PLANTS/PEOPLE/AND ENVIRONMENTAL QUALITY. Gary O. Robinette. U. S. Department of the Interior, National Park Service, 1972. 137 pages.
Describes environmentally responsive landscape design and planting, as well as the environmental aspects of site planning and organization.

THE RSVP CYCLES/CREATIVE PROCESSES IN THE HUMAN ENVIRONMENT. Lawrence Halprin. George Braziller, Inc., 1969. 207 pages.
> Presents design analysis in response to human needs and the environment.

SITE PLANNING. Kevin Lynch. Massachusetts Institute of Technology, 1962. 384 pages.
> Illustrates principles of climatic responsive site planning and site oriented building organization.

geothermal

LOS ALAMOS DRY GEOTHERMAL SOURCE DEMONSTRATION PROJECT. Morton C. Smith, Donald W. Brown and Roland A. Pettitt. Los Alamos Scientific Laboratory of the University of California. 1975.

UTILIZATION OF THERMAL ENERGY AT OREGON INSTITUTE OF TECHNOLOGY, KLAMATH FALLS, OREGON. Oregon Institute of Technology, 1974.

UTILISATION OF GEOTHERMAL ENERGY IN ROTORUA, NEW ZEALAND. W. Burrows. 1974.

GEOTHERMAL DRILLING IN KLAMATH FALLS, OREGON. David M. Storey, E. E. Storey Well Drilling, 1974.
> Several articles on geothermal applications in different parts of the world.

DRY HOT ROCK PROJECT. Francis G. West. Los Alamos Scientific Laboratory of the University of California.

ENERGY FROM DRY GEOTHERMAL RESERVOIRS. M. C. Smith and R. D. McFarland. Los Alamos Scientific Laboratory of the University of California.

FAULT ACTIVITY AND SEISMICITY NEAR THE LOS ALAMOS SCIENTIFIC LABORATORY GEOTHERMAL TEST SITE, JEMEZ MOUNTAINS, NEW MEXICO. David B. Slemmons. Los Alamos Scientific Laboratory of the University of California. 1975.

MAN-MADE GEOTHERMAL RESERVOIRS. Morton C. Smith, R. Lee Aamodt, Robert M. Potter and Donald W. Brown. Los Alamos Scientific Laboratory of the University of California. 1975.

GEOTHERMAL RESERVOIR MODELS – CRACK PLANE MODEL. Robert D. McFarland. Los Alamos Scientific Laboratory of the University of California. 1975.
> Articles from Los Alamos Scientific Laboratory on their experiments on hot rock geothermal exploration.

GEOTHERMAL ENERGY. Resources, Production Stimulation. Edited by Paul Kruger and Carol Otte. Stanford University Press, 1973. 360 pages.
> A collection of 18 articles on the use and exploration of geothermal resources.

GEOTHERMAL ENERGY. Review of research and development. Edited by H. Christopher H. Armstead. The Unesco Press, 1974. 186 pages.
> Covers the basic theories and exploration techniques of geothermal resources.

solar

HERE COMES THE SUN 1981. Joint Venture and Friends. AIA Research Corporation, 1975. 97 pages.
> Book covering the principles of solar collection and its design implications in multi-family housing.

INNOVATION IN SOLAR THERMAL HOUSE DESIGN. Donald Watson. AIA Research Corporation, 1975. Approximately 50 pages.
> Covers the development of concepts for solar thermal design in a climatic responsive house.

NATIONAL PLAN FOR SOLAR HEATING AND COOLING. Interim Report, ERDA, Division of Solar Energy, 1975. 119 pages.
> Explanation of possible residential and commercial demonstration projects.

PROCEEDINGS/WORLD SYMPOSIUM ON APPLIED SOLAR ENERGY. Stanford Research Institute, 1956. 304 pages.
> Interesting book on applications of solar energy in the 1950's and earlier.

SOLARIA. On the Threshold of Environmental Renaissance. Edmund Scientific, Bob and Nancy Homan, Dr. Harry Thomason, Malcolm B. Wells. Edmund Scientific, 1975. 57 pages + drawings.
> Covers the background, design, and construction of the Solaria, solar earth-covered house.

SOLAR ENERGY AS A NATIONAL ENERGY RESOURCE. NSF/NASA Solar Energy Panel, 1972. 85 pages.
> Covers solar energy applications as well as other forms of natural energy.

SOLAR COOLING FOR BUILDINGS. Workshop Proceedings. Semi-annual meeting of ASHRAE. February 6-8, 1974. 231 pages.
> Abstracts and presentations on current research and development of systems to cool buildings using solar energy.

SOLAR ENERGY FOR MAN. B. J. Brinkworth. Halsted Press, John Wiley & Sons, Inc., 1972. 251 pages.
   Review of basic energy principles, applications, and methods for solar energy collections.

SOLAR ENERGY IN BUILDING DESIGN. Bruce Anderson. Total Environmental Action, 1975. Approximately 300 pages.
   A manuscript of an unreleased book written in 1975. Covers many aspects of solar energy and its use in society.

SOLAR ENERGY AND SHELTER DESIGN. Bruce Anderson. Total Environmental Action, 1973. 145 pages.
   A masters thesis on the use of solar energy in buildings.

SOLAR ENERGY – TECHNOLOGY AND APPLICATIONS. J. Richard Williams. Ann Arbor Science Publishers, Inc., 1974. 128 pages.
   Includes techniques and methods for collecting solar energy.

SOLAR HEATED BUILDINGS – A BRIEF SURVEY. W. A. Shurcliff. November, 1975. 172 pages.
   Lists solar heated buildings in the United States. Updated editions come out at two to six month intervals.

Performance of a residential SOLAR HEATING AND COOLING SYSTEM. G.O.G. Lof and D. S. Ward. Annual Progress Report 1974-75, Solar Energy Applications Laboratory, Colorado State University. 56 pages.
   An analysis of the NSF/CSU house in Fort Collins, Colorado.

SOLAR HEATED HOUSES FOR NEW ENGLAND AND OTHER NORTH TEMPERATURE CLIMATES. Mass Design. AIA Research Corporation, 1975. 68 pages.
   Examples of solar houses suited for the New England area. Also included is a sample computer program to determine the availability of and the requirement for solar energy in houses.

SOLAR HEATING AND COOLING. Engineering, Practical Design and Economics. Jan F. Kreider and Frank Kreith. Hemisphere Publishing Corporation, McGraw Hill, 1975. 342 pages.
   A definitive text on solar energy, heat transfer principles, and solar heating and cooling of buildings.

SOLAR HEATING HANDBOOK FOR LOS ALAMOS. D. Balcomb, J. C. Hedstrom, S. W. Moore, and B. T. Rogers. Los Alamos Scientific Laboratory of the University of California, 1975. 71 pages.
   Performance calculations on solar test facilities at LASL. Examples of solar heated buildings are included.

SOLAR HOUSE PLANS. Harry E. Thomanson and Harry Jack Lee Thomanson, Jr. Edmund Scientific Company, 1972. 18 pages.
   Explanation plus fold-out drawings covering construction and installation of collectors on the Thomanson house.

SOLAR USE NOW – A RESOURCE FOR THE PEOPLE. Extended Abstracts, ISES Conference, Los Angeles, 1975. 540 pages.
   Abstracts of papers presented at the International Solar Energy Society Conference, August, 1975, in Los Angeles.

SOLAR ORIENTED ARCHITECTURE. College of Architecture, Arizona State University. AIA Research Corporation, 1975. 142 pages.
   A survey of solar heated and/or cooled buildings in the U.S. Included are basic principles of solar energy.

THE SOLAR RESOURCE. 14 Articles on Energy from the Sun. Environmental Action Reprints, 1974. Approximately 40 pages.
   Reprints from other sources of articles on wind and solar energy.

SOLAR THERMAL ENERGY UTILIZATION. Volumes I and II. Energy Information Center of Technology Application. University of New Mexico, 1974. Approximately 1500 pages.
   Index containing thousands of solar related articles and books.

THE SUN IN ART. Walter Herdeg. The Graphis Press, 1968. 156 pages.
   Examples of man's use of the sun in his art throughout the ages.

SUN ANGLE CALCULATOR. Libbey-Owens-Ford Company, 1975.
   A device to determine the position of the sun during the year or day at many different latitudes.

TECHNICAL PROGRAM AND ABSTRACTS. International Solar Energy Society, U.S. Section Annual Meeting, 1974. 221 pages.
   A collection of articles and notes on the ISES Conference in August, 1974.

wind

AN INTRODUCTION TO THE USE OF WIND. Douglas R. Coonley. Total Environmental Action, 1975. Approximately 80 pages.
   Presents the basic principles of wind energy.

WIND ENERGY CONVERSION SYSTEMS. Workshop Proceedings. NSF/NASA, 1973. 258 pages.
   Articles and presentations at the NSF/NASA Wind Energy Conference.

WIND MACHINES. Frank Eldridge, Mitre Corporation, 1975. 77 pages.
   An analysis and history of many types of wind machines. Also included is a presentation of vortex type wind machines and large scale electrical power generation using wind energy.

WINDMILLS & WATERMILLS. John Reynolds. Praeger Publishers, Inc., 1970. 196 pages.
   A collection of descriptions, photographs and excellent illustrations of windmills and watermills in Europe and the USA.

notes

ABSORPTANCE – A ratio of the radiation absorbed by a surface to the radiation incident on that surface.

ABSORPTION REFRIGERATION – A cooling system that uses heat as its primary source of energy and evaporation as the cooling means.

AIR CHANGE – The replacement of the air contained within an enclosed space within a given period of time.

AIR CONDITIONING – The process of treating air to control its temperature, humidity, flow, cleanliness, and odor.

ALTERNATING CURRENT (AC) – Electrical current in which the direction of electron flow is reversed at regular intervals. In the USA, 60 hertz (cycles per second) is the standard frequency. In Europe, 50 hertz is standard.

AMBIENT TEMPERATURE – The temperature of the outside air or air surrounding a space or building.

AMPERE (AMP) – A measure of the quantity of electric current flowing in a circuit. One volt applied across a resistance of one ohm will cause one amp to flow.

AUXILIARY FURNACE – A supplementary heating unit used to provide heat to a space when its primary heat source cannot do so adequately.

BARREL – A measure of a quantity of fluid, usually equal to 42 U.S. gallons, 5.6 cubic feet, or 159 liters.

BATTERY – A device to store electrical energy as chemical potential energy. Within limits, most can be repeatedly charged and discharged.

BERM – A mound or small hill of earth, man-made.

BIOCONVERSION – Conversion of solar energy to fuel by the natural process of photosynthesis.

BIOFUELS – Renewable fuels and energy sources derived from organic material (wood, methane, etc.).

BIOSPHERE – The zone of air, land, and water, above and below the earth's surface, which is occupied by plants and animals.

BLACKOUT – A complete shutoff of electrical energy from a power generating source due to overloads, power shortages, or outage caused by equipment or transmission breakdown.

BOILER – A device used to heat water or to produce steam for space heating or other uses, including power generation.

BRITISH THERMAL UNIT (Btu) – The amount of heat necessary to raise the temperature of one pound of water one degree Fahrenheit.

BROWNOUT – A reduction in line voltage planned to alleviate overloads on power generating equipment. Reduced voltage diminishes the brightness of incandescent lamps.

BUILDING ENVELOPE – Exterior components of construction which enclose an interior space.

CALORIE – The amount of heat necessary to raise the temperature of one gram of water one degree Celsius.

CHANGE OF STATE (Phase change) – The change from the solid, liquid, or gaseous state to either of the other two.

COEFFICIENT OF PERFORMANCE (COP) – The ratio of the energy output of a device, such as a heat pump, to the energy input.

COLLECTOR TILT – The angle at which a solar collector is inclined with respect to a horizontal plane.

COLOR RENDITION – The effect of a light source on an object's perceived color.

COLOR TEMPERATURE – The measurement in degrees Kelvin of the color output of a light source.

COMFORT ZONE – The range of temperatures and humidities over which the majority of adults feel comfortable under normal living and working conditions.

CONCENTRATOR – A device used to intensify the solar radiation striking a surface.

CONDENSATION – The process of changing a vapor into a liquid by the extraction of heat.

CONDENSER – A component of a system in which a working fluid undergoes a change of state from gas to liquid by the rejection of latent heat to a cooling medium, producing a heating effect.

COOLING LOAD – The amount of heat which must be removed from a building to maintain a comfortable temperature, measured in Btu/hr or tons of air conditioning. (1 ton = 12,000 Btu/hour).

COOLING POND – A body of water which dissipates heat by evaporation, convection, and radiation.

CROOKE'S RADIOMETER – A partially evacuated hollow glass sphere containing vanes which are black on one side and silvered on the other side. These spin in the presence of thermal radiation due to differing rates of radiation absorption between the silvered and the blackened surfaces.

CRUDE OIL – The natural mixture of liquid hydrocarbons extracted as petroleum from under the earth's surface. Also may include oil extracted from tar sands and oil shale.

DESIGN CONDITIONS – Selected indoor and outdoor wet bulb and dry bulb temperatures for a specific location which determine the maximum heating and cooling loads of a building.

DEGREE DAY, HEATING – A unit used to estimate the heating fuel consumption and the nominal heating load of a building. For any one day, when the mean temperature is less than 65 F, there exist as many degree days as there are Fahrenheit degrees difference in temperature between the mean temperature for the day and 65 F. The sum of the degree days constitutes the annual degree day heating requirement.

DIRECT CURRENT (DC) – Electrical current in which the direction of electron flow never changes.

DISABILITY GLARE – Reflected light which impairs a viewer's ability to accurately discern an object, perceive color, or to read printed matter.

DRY BULB TEMPERATURE – The local air temperature as indicated by a dry temperature measuring sensor (the one with which we are most familiar).

DUCT – A conduit or tube through which air or other gases flow.

ECONOMIZER CYCLE – A cooling mode which uses cool outdoor air to offset heat gains in building rather than using an energy consuming cooling device.

EFFICIENCY, THERMAL – The ratio of the useful heat at the point of use to the thermal energy input for a designated time period, expressed as percent.

EFFLUENT – Discharged wastes suspended in liquid or gas.

ELECTRIC DEMAND LIMITER – A device that selectively switches off electrical equipment whenever total electrical demand rises beyond a predetermined level.

ELECTRIC RESISTANCE HEAT – The conversion of electrical energy to heat energy, using an electrical resistor. May be applied to space heating or water heating. 1KWh = 3,413 Btu.

ELECTRICITY – The interaction between particles of positive and negative change. Utilized as a flow of electrons, producing electric current.

ELECTROMAGNETIC SPECTRUM – The entire range of wavelengths of electromagnetic radiation, extending from gamma rays to the longest radio waves.

EMITTANCE – A ratio of the amount of energy radiated by a surface to the energy striking the surface.

ENERGY – The capacity for doing work; taking a number of forms which may be transformed from one into another, such as thermal (heat), mechanical (work), electrical, and chemical; in customary units, measured in kilowatt hours (KWh) or British thermal units (Btu); in SI units, measured in joules (J), where 1 joule = 1 watt-second.

   DIRECT ENERGY – Energy used in its most immediate form, that is, natural gas, electricity, oil.

   INDIRECT ENERGY – Energy that is converted into goods which are then consumed. Examples: food through photosynthesis, fibers, plastics, chemicals.

   NET ENERGY – The energy remainder or deficit after the energy costs of extracting, concentrating, and distribution are subtracted.

   NET RESERVES – An estimate of the net energy that can be delivered from a given energy resource.

ENERGY STORAGE – The ability to retain energy by converting it to a form (gravitational potential, chemical, etc.) from which it can be retrieved for useful purposes.

ETHYL ALCOHOL – A colorless, volatile, flammable liquid (C_2H_5OH) which is produced by the fermentation of grains and fruits.

EUTECTIC SALT – A combination of two (or more) mutually soluble materials, requiring a large amount of heat to melt or an equally large amount of heat to solidify, which melts or freezes at constant temperature and with constant composition. The phase change temperature is the lowest achievable with these materials.

EVAPORATOR – A component of a system in which a working fluid undergoes a change of state from liquid to gas, taking on latent heat and thereby providing a cooling effect.

FAN COIL – A heating or cooling device that forces air through heating or cooling coils.

FISSION – The splitting of an atomic nucleus, releasing large amounts of energy. Usually the uranium 235 atom is split, producing heat for the generation of electricity with steam.

FLOW RATE – The volume or weight per unit time of a fluid flowing through an opening or duct.

FLUE – The exhaust channel through which gas and fumes produced by combustion exit a building.

FOOTCANDLE – Unit of measurement of the intensity of light on a surface that is everywhere one foot from a uniform point source of light from a standard (sperm whale oil) candle and equal to one lumen per square foot.

FOSSIL FUEL – A class of organic compounds, formed by the decay of matter under the influence of heat and pressure over millions of years, which when burned liberate large quantities of energy, exemplified by coal, oil, and natural gas.

FROSTLINE – The depth of frost penetration in the earth. This depth varies from one geographic location to another.

FUEL – Any substance that can be burned to produce heat.

FUEL CELL – A device in which hydrogen and oxygen are combined in an electro-chemical reaction to generate electricity and produce water as a by-product.

FUSION (NUCLEAR) – The release of energy by the formation of a heavier nucleus from two lighter ones.

GEOTHERMAL ENERGY – Heat energy contained in large underground reservoirs of steam and hot water, produced by molten material from the earth's interior.

HEAT EXCHANGER – A device used to transfer heat from one temperature level to another.

HEAT GAIN – An increase in the amount of heat contained in a space, resulting from solar radiation and the heat given off by people, lights, equipment, machinery, and other sources.

HEAT LOSS – A decrease in the amount of heat contained in a space, resulting from heat flow through walls, windows, roof, and other components of the building envelope and from the infiltration of cold outdoor air.

HEAT PIPE – A closed pipe containing a liquid and a wick which will transfer heat from one end to the other end without any input of work.

HEAT PUMP – A reversible refrigeration system that delivers more heat energy to the end use than is input to the compressor. The additional energy input results from the absorption of heat from a low temperature source.

HEAT SINK – A body or substance which is capable of accepting or rejecting heat.

HEAT WHEEL – A device used in ventilating systems which tends to bring incoming air into thermal equilibrium with exiting air. As a result, hot summer air is cooled and cold winter air is warmed.

HELIODON – A device used to simulate the effect the sun's position has on models of buildings and other objects, primarily used to conduct shadowing studies.

HELIOSTAT – An instrument consisting of a mirror mounted on an axis, mechanically rotated to steadily reflect the sun in one direction.

HELIOTROPISM – Property of being able to follow the sun's apparent motion across the sky.

HORSEPOWER – A standard unit of power equal to 746 watts, 2545 Btu per hour, or 550 foot-pounds per second.

HUMIDISTAT – A device to continuously monitor the humidity in a space and activate the humidity control equipment to insure that the humidity is maintained at a pre-set level.

HVAC – Abbreviation for heating, ventilating, and air conditioning.

HYDROELECTRIC PLANT – An electrical power generating plant in which the kinetic energy of water is converted to electrical energy using a turbine generator.

HYDRONIC HEATING – A heating system in which liquid is used for heat transport.

ILLUMINATION – Light provided to interior and exterior spaces.

INFILTRATION – The uncontrolled flow of air into a building through cracks, openings, doors, or other areas which allow air to penetrate.

INSOLATION – Incident Solar Radiation – The amount of solar radiation striking a surface during a specified period of time.

KILOWATT (kW) – A unit of power equal to 1000 watts.

KILOWATT HOURS (kWh) – A unit of work equal to the consumption of 1000 watts in one hour.

LANGLEY – A measure of irradiation in terms of langleys per minute, where one langley equals 1 calorie per square centimeter. Named in honor of American astronomer Samuel P. Langley.

LATENT HEAT – A change in heat content due to a change in state that occurs without a corresponding change in temperature.

LUMEN – A unit measure of the light output of a lamp, where one lumen provides an intensity of one footcandle at a distance of one foot from the light source.

LUX – A measure of light intensity on a plane, denoting lumens per square meter.

MEGAWATT – A unit of power equal to one million watts or one thousand kilowatts.

METHANE – A colorless, odorless, flammable gaseous hydrocarbon (CH_4), which is the product of the decomposition of organic matter. It is the major component of natural gas.

PEAK LOAD – The maximum energy demand placed on a system.

PHOTOLYSIS – Chemical decomposition caused by the action of solar radiant energy.

PHOTOSYNTHESIS – The production of chemical compounds in plants using solar radiant energy, specifically the production of carbohydrates by plants containing chlorophyll using sunlight, carbon dioxide, and water.

PLENUM – A compartment for the passage and distribution of air.

POWER – The rate at which work is performed or energy expended.

QUAD – 10^{15} Btu (one quadrillion Btu).

RAW ENERGY SOURCE - An original unrefined source of energy.

RECOVERED ENERGY - Energy utilized which would otherwise be wasted.

RECYCLE - The process of recovering resources from waste material for reprocessing and reuse.

REFINERY - A chemical processing plant in which crude oil is separated into more useful hydrocarbon compounds.

REFLECTANCE - The ratio of the amount of radiation reflected by a surface to the amount of radiation incident on the surface.

SELECTIVE SURFACE - Pertaining to solar collectors, the surface produced by the application of a coating to an absorber plate, with spectral selective properties, which maximizes the absorption of incoming solar radiation (0.3 to 3.0 micron range) and emits much less radiation (3.0 to 30.0 micron range) than an ordinary black surface at this same temperature would emit.

SENSIBLE HEAT - Heat that produces a temperature change.

SKYSHAFT - A multi-chambered plexiglass device which penetrates a roof used to provide natural interior illumination with a minimum of heat loss or gain.

SOLAR CELL - Also, photovoltaic cell. A device employing crystals (silicon, for example) which, when exposed to solar radiation, generates an electric current.

SOLAR COOKER - A device for cooking which uses the sun as an energy source.

SOLAR ENERGY - Energy in the form of electromagnetic radiation received from the sun.

SOLAR POWER FARM - An installation for generating electricity on a large scale using solar energy, consisting of an array of solar collectors, steam or gas turbines, and electrical generators.

SPACE HEATING - Interior heating of a building or room.

SUN TIME - Time of day at a specific location as determined by the position of the sun.

SUN TRACKING - The ability to follow the apparent motion of the sun across the sky.

THERM - A quantity of heat equal to 100,000 Btu.

THERMAL TRANSMISSION - The passage of heat through a material.

THERMOMETER - An instrument for measuring temperature.

THERMOSTAT - A device to continuously monitor the temperature in a space and activate temperature control equipment to insure that the temperature in the space is maintained at a pre-set level.

TON (of air conditioning) - The thermal refrigeration energy required to create one ton of ice (2000 pounds) in one day. Equals 12,000 Btu/hour.

TORQUE - A turning or twisting force.

U-VALUE - A coefficient which indicates the time rate at which energy (Btu/hour) passes through a component for every degree (Fahrenheit) of temperature difference between one side and the other under steady state conditions.

VAPOR BARRIER - A component of construction which is impervious to the flow of moisture, used to prevent moisture travel to a point where it may condense.

VENT - Any penetration through the building envelope, which is specifically designed for the flow of air into or from the building.

VENTILATION AIR - Outside air that is intentionally caused to enter an interior space.

WEATHERSTRIPPING - Foam, metal or rubber strips used to form an air-tight seal around windows, doors, or openings to reduce infiltration.

WET BULB TEMPERATURE - The local air temperature as indicated by a wet temperature measuring sensor.

WIND ENERGY - The kinetic energy of air motion over the earth's surface caused by the sun's heating of the atmosphere.

WIND MACHINE - Any one of a number of devices used to convert the kinetic energy of wind to another form of energy for useful purposes.

WORK - The transfer of energy from one physical system to another. In mechanics, the transfer of energy to a body by the application of force.

notes